T0205513

Lecture Notes in Statistics

Volume 224

Series Editors

Peter Diggle, Department of Mathematics, Lancaster University, Lancaster, UK

Scott Zeger, Baltimore, MD, USA

Ursula Gather, Dortmund, Germany

Peter Bühlmann, Seminar for Statistics, ETH Zürich, Zürich, Switzerland

Lecture Notes in Statistics (LNS) includes research work on topics that are more specialized than volumes in Springer Series in Statistics (SSS).

The series editors are currently Peter Bühlmann, Peter Diggle, Ursula Gather, and Scott Zeger. Peter Bickel, Ingram Olkin, and Stephen Fienberg were editors of the series for many years.

Manfred Deistler · Wolfgang Scherrer

Time Series Models

 Springer

Manfred Deistler
Institute of Statistics and Mathematical
Methods in Economics
TU Wien
Vienna, Austria

Institute for Statistics and Mathematics
WU Wien
Vienna, Austria

Wolfgang Scherrer
Institute of Statistics and Mathematical
Methods in Economics
TU Wien
Vienna, Austria

ISSN 0930-0325   ISSN 2197-7186   (electronic)
Lecture Notes in Statistics
ISBN 978-3-031-13212-4   ISBN 978-3-031-13213-1   (eBook)
https://doi.org/10.1007/978-3-031-13213-1

Mathematics Subject Classification: 60G10, 62M10, 62M15, 62M20, 93E12, 62P20, 62P30, 62H25

Translation from the German language edition: "Modelle der Zeitreihenanalyse" by Manfred Deistler and
Wolfgang Scherrer, © Springer International Publishing AG 2018. Published by Birkhäuser. All Rights
Reserved.

© Springer Nature Switzerland AG 2022
This work is subject to copyright. All rights are reserved by the Publisher, whether the whole or part of
the material is concerned, specifically the rights of translation, reprinting, reuse of illustrations, recitation,
broadcasting, reproduction on microfilms or in any other physical way, and transmission or information
storage and retrieval, electronic adaptation, computer software, or by similar or dissimilar methodology
now known or hereafter developed.
The use of general descriptive names, registered names, trademarks, service marks, etc. in this publication
does not imply, even in the absence of a specific statement, that such names are exempt from the relevant
protective laws and regulations and therefore free for general use.
The publisher, the authors, and the editors are safe to assume that the advice and information in this book
are believed to be true and accurate at the date of publication. Neither the publisher nor the authors or
the editors give a warranty, expressed or implied, with respect to the material contained herein or for any
errors or omissions that may have been made. The publisher remains neutral with regard to jurisdictional
claims in published maps and institutional affiliations.

This Springer imprint is published by the registered company Springer Nature Switzerland AG
The registered company address is: Gewerbestrasse 11, 6330 Cham, Switzerland

We both want to thank our families for their understanding and support.

Preface

A time series consists of observations or measurements ordered in time. Thereby the information is contained not only in the observed values, but also in their ordering in time. This is contrary to the case of "classical" i.i.d statistics, where a permutation of the data leaves the results unchanged. Time series analysis is concerned with the extraction of information from time series data and hence is a part of statistics. As in statistics in general, the focus of time series analysis is on data-driven modeling, where the data are assumed to be generated by a stochastic model. In time series analysis, these models are naturally often dynamic, i.e. they describe the evolution in time of the variables considered. The models obtained in this way can be used for instance for analysis, forecasting, filtering or control. However, data-driven modeling is not the only branch of time series analysis; for instance, (not model-based) denoising of signals or extraction of features, such as hidden cycles, are important parts of time series analysis as well.

The history of time series analysis, or to be more precise, the development and application of methods of time series analysis (which transcends analysis by the "naked eye"), goes back to the turn of the eighteenth to the nineteenth century and was triggered by the question, whether deviations of the orbits of planets from the elliptic form are detectable. The deviations from elliptic orbits described by Kepler are a consequence of the fact that the planets and the sun form a multi-body system rather than a two-body system. The so-called periodogram was developed in this context. The periodogram was already in use for the analysis of business cycle data in the nineteenth century. Moving average (MA) and autoregressive (AR) processes were introduced by G. U. Yule in the 1920s. The theory of (univariate) stationary processes was pioneered in the 1930s and 1940s by A. N. Kolmogorov, H. Cramér, N. Wiener and K. Karhunen. The multivariate case was developed in particular by Y. A. Rozanov (see Rozanov (1967) and the references given therein). This theory still is an essential foundation for the analysis of time series.

The development of time series analysis took place in a number of different fields such as econometrics, system and control theory, signal processing and statistics. The main advancements in time series analysis over the last 75 years were as follows:

- The analysis of the problem of identifiability and the maximum likelihood estimation in multivariate, "structural" ARX systems pioneered by the Cowles Commission.
- The development of non-parametric spectral estimation by J. Tukey, in particular.
- The analysis of AR and ARMA systems (ARX and ARMAX systems, respectively) in particular by T. W. Anderson and E. J. Hannan. The book Box and Jenkins (1976) triggered the wide-spread use of these systems in various fields of applications. Subsequently, a generalization to the multivariate case has been carried out, as documented in the books Caines (1988), Hannan and Deistler (2012), Ljung (1999), Lütkepohl (1993) and Reinsel (1997).
- The analysis of state-space systems and the related Kalman filter, mainly by R. E. Kalman.
- The introduction and analysis of methods for order estimation, e.g. by H. Akaike and J. Rissanen.
- Over the last thirty years, the analysis of integrated and cointegrated processes, i.e. of a particular class of non-stationary processes, has obtained great attention in econometrics. Important contributions in this context are due to C. W. J. Granger, R. F. Engle, P. C. B. Phillips and S. Johansen.
- Models for forecasting conditional variances for risk assessment for financial time series, such as ARCH or GARCH models, were introduced by R. F. Engle.
- Dynamic factor models have become popular in the last decades, since they can deal with high-dimensional time series.
- The huge area of non-linear time series models and of their estimation (see, e.g. the book Pötscher and Prucha (1997)) has undergone a rapid development over the last 25 years.
- Recently neural nets have been successfully used for the analysis and prediction of time series. See e.g. the book Goodfellow et al. (2016).

This book by far does not cover all the important parts of time series analysis. We focus on *models* for time series and in particular on the most important class of linear models. We discuss in detail weakly stationary processes as well as important subclasses, including AR and ARMA processes. Our analysis focuses on the multivariate case. Both the "linear" theory of weakly stationary processes as well as linear dynamical systems still form the main part of the foundations of time series analysis, despite the fact that non-stationarity and non-linearity are of major importance. It is a specific feature of time series analysis—as opposed to other areas of statistics—that an accurate analysis of the models is important for statistical analysis in the narrow sense.

We aim to provide readers with a solid basis that allows them to understand large parts of the current literature in the field of time series analysis. In a certain sense, this book is meant to convey essential core knowledge. As compared to other textbooks in this area, this book focuses on the very core of the structure theory for multivariate linear time series models. We try to give a comprehensible and detailed presentation of this material.

Primarily, this book addresses mathematicians and advanced students of mathematics. However, we believe that it is equally accessible and useful for researchers in the fields of econometrics, financial mathematics, control engineering or signal processing. Knowledge of measure- and probability theory and of linear algebra are prerequisites. Moreover, the reader should be familiar with the basics of functional analysis (theory of Hilbert spaces) and of complex analysis.

The subject matter is broken down into several chapters: Chap. 1 provides basic definitions of (weakly) stationary processes, their embedding in the Hilbert space of square integrable random variables, as well as the definition of corresponding covariance functions; the latter, for many problems, contain essential information about the underlying stationary process. Important classes of models for stationary processes are discussed at the end of this chapter.

Chapter 2 deals with the linear least squares prediction of stationary processes. The central result here is Wold's decomposition, which gives substantial insight into the structure of general stationary processes.

While the description of stationary processes in Chaps. 1 and 2 is done in the time domain, Chap. 3 covers the frequency domain. Key results are the spectral representation of covariance functions and of stationary processes, both being Fourier representations. From the spectral representations, we obtain the spectral distribution function and the spectral density, respectively, which contain the same information about the underlying process as the covariance function. Due to these Fourier representations, linear dynamical transformations of stationary processes correspond to a multiplication of functions (in frequency domain) and therefore are often easier to represent and interpret. This chapter is the most challenging one in terms of mathematics and the results will be used in subsequent chapters. However, the remaining chapters are accessible even if the proofs in this chapter have been omitted in the first reading.

The next Chap. 4 describes linear dynamical transformations of stationary processes in the time and frequency domain, as well as the corresponding transformation of the second moments. Such linear transformations are important models for real-world systems and are used to construct classes of stationary processes such as AR and ARMA processes. In this context, we also discuss the solutions of linear stochastic difference equations. At the end of this chapter, we will discuss the Wiener filter, which gives the best approximation, in a mean squares sense, of a stationary process by a linear transformation of a second process.

Chapter 5 deals with AR systems and AR processes, which are the most important model class in time series analysis. This class allows for an arbitrary close approximation of regular processes. Estimation and prediction of these processes are very simple. Beyond the stationary case, AR systems also serve as models for integrated and co-integrated processes, which have gained great importance in econometrics.

Chapter 6 treats ARMA models and ARMA processes. We show that the class of ARMA processes is exactly the class of stationary processes with rational spectral density. As is the case with AR processes, any regular, stationary process can be approximated by an ARMA process with any degree of accuracy. ARMA processes are more flexible, which means that in many cases fewer parameters are needed for approximation. However, the structure of the class of ARMA processes is much more complex as compared to the AR case. There is a so-called problem of identifiability and the relationship between the population second moments of the observations and the ARMA parameters is not given by a linear system of equations, contrary to the AR case. Hence, estimation of ARMA parameters (which is not discussed here) is much more difficult than in the AR case.

Chapter 7 discusses state-space systems, which are of central importance in control engineering. We will show that state-space systems with white noise inputs, like ARMA systems, describe the class of all processes with rational spectra. In this sense, ARMA and state-space systems are equivalent. Under suitable conditions, these two representations lead directly to the Wold representation. In the last section of this chapter, we will discuss the Kalman filter, which is an extremely important algorithm, particularly for the approximation of the unobserved state as well as for prediction and filtering.

Chapter 8 deals with linear models with exogenous variables, in particular the so-called ARX and ARMAX models and the corresponding state-space models. In addition, we discuss the identifiability of models with additional prior information on the parameters in this context.

Chapter 9 presents Granger causality, which is a concept to formalize causal relations or dependencies.

In the subsequent Chap. 10, the dynamic factor models are introduced. Such models are of particular importance for high-dimensional time series, since they avoid the so-called "curse of dimensionality". The basic idea is to decompose the observed variables into a part generated by a "small" number of hidden factors and some "residual" noise.

The last chapter deals with (multivariate) GARCH models which describe the conditional variance and thus are often used to describe risk in finance. This is the only chapter where non-linear time series models are treated.

As mentioned above, important areas are not addressed in this book. As far as linear model classes are concerned, the book does not deal, e.g. with non-stationary processes (with an exception of the integrated case), such as locally stationary processes or with models for functional time series. We do also not consider models for continuous-time observations, which are important for high-frequency financial times series or for applications in systems and control engineering, where often the physical models are in continuous time. We do not treat the large class of non-linear models, with the exception of GARCH models.

This book is restricted to the discussion of model classes and the relation between internal model parameters and the population second moments of the observations (structure theory), which is of great importance for time series analysis. However, it does not cover estimation and inference in the narrow sense. In particular, we will not discuss the estimation of expected values, covariance functions, spectral densities, AR, ARMA or state-space systems. We also do not deal with the large field of Bayesian time series analysis.

We think that time series analysis is a fascinating field with a wide range of applications and a highly nontrivial mathematical theory. Examples of application are the prediction or seasonal adjustment of economic variables, the design of controllers for chemical processes, the transmission and denoising of speech signals, the analysis of signals in radio astronomy or the analysis of electroencephalograms.

This book is an extended translated version of the German book "Modelle der Zeitreihenanalyse" (Deistler and Scherrer (2018), published by Birkhäuser), where in the English version we have added Chaps. 8–11. Some parts of the book are based on lectures we gave at the Vienna University of Technology, at the Institute for Advanced Studies, Vienna (IHS) and at CERGE-EI in Prague. We would also like to thank our colleagues Dietmar Bauer, Marianna Bolla, Philip Gersing, Marco Lippi, Christoph Rust and Otmar Scherzer for their valuable comments. In addition, we also thank the translator Georg Raslagg for his contribution to the translation of the German book.

Vienna, Austria   Manfred Deistler
June 2022   Wolfgang Scherrer

References

G.E.P. Box and G.M. Jenkins. *Time Series Analysis: Forecasting and Control.* Holden-Day, San Francisco, revised edition, 1976.

E.P. Caines. *Linear Stochastic Systems.* John Wiley & Sons, New York, 1988.

M. Deistler and W. Scherrer. *Modelle der Zeitreihenanalyse.* Mathematik Kompakt. Birkhäuser, 2018. Editors: Brokate, M., Hoffmann, K.-H., Kersting, G., Reiss, K., Scherzer, O., Stroth, G., Welzl, E.

I. Goodfellow, Y. Bengio, and A. Courville. *Deep Learning.* MIT Press, 2016. URL http://www.deeplearningbook.org.

E.J. Hannan and M. Deistler. *The Statistical Theory of Linear Systems.* Classics in Applied Mathematics. SIAM, Philadelphia, 2012. Originally published: John Wiley & Sons, New York, 1988.

L. Ljung. *System Identification: Theory for the User.* Prentice Hall, second edition, 1999.

H. Lütkepohl. *Introduction to Multiple Time Series Analysis.* Springer, Berlin, 2nd edition, 1993.

B.M. Pötscher and I. Prucha. *Dynamic Nonlinear Econometrics Models.* Springer, Berlin-Heidelberg, 1997.

G.C. Reinsel. *Elements of Multivariate Time Series Analysis.* Springer, 1997.

Y.A. Rozanov. *Stationary Random Processes.* Holden-Day, San Francisco, 1967.

Contents

Time Series and Stationary Processes

1

This chapter introduces basic concepts such as time series, stationary process and covariance function. Subsequently, the time domain of a stationary process, which is a subspace of the Hilbert space $\mathbb{L}_2$ of square integrable random variables, is presented. The last section describes classes of stationary processes and examples of non-stationary processes. The concept of weakly stationary processes goes back to Khinchin.[1] The book Doob (1953)[2] contains a detailed description of the one-dimensional case. The properties of covariance functions were described by Khinchin and Wold,[3] and for the multivariate case by Cramér.[4] Textbooks on this topic are, e.g. Brockwell and Davis (1991), Bolla and Szabados (2021) and Pourahmadi (2001).

1.1   Data Structure: Time Series

A time series consists of a finite number of measured values ordered in time

$$x_{t_1}, x_{t_2}, \ldots, x_{t_T}; \ x_{t_k} \in \mathbb{R}^n, \ k = 1, 2, \ldots, T,$$

where $t_1 < t_2 < \cdots < t_T$. A time series is called *scalar* or *univariate*, if $n = 1$, i.e. there is one measurement for every time stamp. If $n > 1$ observations are present

---

[1] Aleksandr Yakovlevich Khinchin (1894–1959). Russian mathematician. His main field was stochastics.

[2] Joseph L. Doob (1910–2004). US-American mathematician. His research interests included analysis and probability theory (with particular emphasis on stochastic processes).

[3] Herman Wold (1908–1992). Swedish statistician. He investigated stationary processes and developed the Wold decomposition, which was named after him.

[4] Harald Cramér (1893–1985). Swedish mathematician and statistician. Herman Wold's doctorate supervisor.

© Springer Nature Switzerland AG 2022
M. Deistler and W. Scherrer, *Time Series Models*, Lecture Notes in Statistics 224,
https://doi.org/10.1007/978-3-031-13213-1_1

at any time, we speak of *multivariate* time series. It should be emphasized that information in a time series is not only contained in the individual measured values, but also in their order in time. In many cases, time series analysis is about analyzing the relation of the measured values at different points in time.

This book focuses only on equally spaced time series, where $t_k = t_0 + \Delta k$ for the measurement time points. The variable $\frac{1}{\Delta}$ is called sampling rate. Without loss of generality, we set $t_0 = 0$ and $\Delta = 1$ and then write the time series as

$$x_t, \quad t = 1, \ldots, T.$$

Example (macroeconomic time series)

Nominal gross domestic product in Austria and Germany, quarterly data (1988Q1-2013Q3), $T = 103$ measured in million EUR, source: Eurostat.

Period	Germany	Austria
1988Q1	248115.6	25681.6
1988Q2	257852.0	27531.2
1988Q3	266906.4	29184.7
1988Q4	286152.0	29303.6
1989Q1	266020.3	27801.9
$\vdots$	$\vdots$	$\vdots$
2013Q3	703580.0	80082.2

In this case, equidistance is an idealization because not all quarters consist of the same number of days.

1.2   Stationary Processes and Covariance Function

Definition 1.1 (*stochastic process*) A *stochastic process* $(X_t \mid t \in \mathbb{T})$ is a family of random vectors (random variables), which are defined on a (common) probability space $(\Omega, \mathcal{A}, \mathbf{P})$. In most cases, we consider real-valued random vectors, i.e. $X_t \colon \Omega \to \mathbb{R}^n$, $X_t = (X_{1t}, \ldots, X_{nt})'$. However, in connection with spectral representation, we will also discuss $\mathbb{C}^n$-valued random vectors.

The index set $\mathbb{T}$ is usually interpreted as time. If this set is countable (e.g. $\mathbb{T} = \mathbb{Z}$ or $\mathbb{T} = \mathbb{N}_0$), this process is called *discrete time*. *Continuous-time* processes are defined on $\mathbb{T} = \mathbb{R}$, $\mathbb{T} = \mathbb{R}_+$ or on intervals such as $\mathbb{T} = [0, 1]$. In this book, we will look almost exclusively at processes defined on $\mathbb{T} = \mathbb{Z}$. In this case, one usually only writes (X_t) instead of $(X_t \mid t \in \mathbb{Z})$. The process is *scalar* or *univariate* for $n = 1$ and *multivariate* otherwise.

Definition 1.2 (*trajectory*) The function $t \longmapsto X_t(\omega)$, for fixed $\omega \in \Omega$, is called *trajectory* or *path*.

A stochastic process can be interpreted, as defined above, as

- a family of random vectors (random variables)

$$t \longmapsto \begin{pmatrix} X_t(\cdot): & \Omega \longrightarrow \mathbb{R}^n \\ & \omega \longmapsto X_t(\omega) \end{pmatrix}$$

- or alternatively as a "random function"

$$\omega \longmapsto \begin{pmatrix} X.(\omega): & \mathbb{T} \longrightarrow \mathbb{R}^n \\ & t \longmapsto X_t(\omega) \end{pmatrix}.$$

We will primarily use the first approach.

If $\mathbf{E}|X_{it}| < \infty$ for all $i = 1, \ldots, n$ and $t \in \mathbb{Z}$, then

$$\mu: \mathbb{Z} \longrightarrow \mathbb{R}^n$$
$$t \longmapsto \mu(t) = \mathbf{E}X_t$$

is called the *mean value function* of the process (X_t). Here (and in the following) $\mathbf{E}$ denotes expectation. Furthermore, we use $\mathbf{Cov}(X, Y)$ for the covariance (matrix) between the random vectors X, Y and $\mathbf{Var}(X)$ for the variance (matrix) of X. If the process $(X_t \mid t \in \mathbb{Z})$ is square integrable, i.e. if $\mathbf{E}X_{it}^2 < \infty$ applies to all $i = 1, \ldots, n$ and $t \in \mathbb{Z}$, then one defines

$$\gamma: \mathbb{Z} \times \mathbb{Z} \longrightarrow \mathbb{R}^{n \times n}$$
$$(t, s) \longmapsto \gamma(t, s) = \mathbf{Cov}(X_t, X_s) = \mathbf{E}(X_t - \mu(t))(X_s - \mu(s))'$$

as the *covariance function* of the process (X_t).

Here, stochastic processes are mostly *models* for time series. In other words, one (mostly) assumes that the observed time series was "generated" by an underlying stochastic process:

$$(x_t = X_t(\omega) \in \mathbb{R}^n \mid t = 1, \ldots, T).$$

Thus, the observed time series is a finite part of a trajectory of the *data-generating process* (DGP). This allows for statistical conclusions from the time series about the properties of the underlying stochastic process.

In the following notation, we will *no longer* distinguish between random variables or random vectors and realizations, i.e. x_t, for example, stands for a random vector as well as for a realization of this random vector.

Definition 1.3 (*stationary process*)

- A stochastic process $(x_t \mid t \in \mathbb{Z})$ is *strictly stationary*, if the joint distribution of $(x'_{t_1+k}, \ldots, x'_{t_s+k})'$ is independent of k for all finite subsets $\{t_1, \ldots, t_s\} \subset \mathbb{Z}, s > 0$ and for all $k \in \mathbb{Z}$.
- A stochastic process $(x_t \mid t \in \mathbb{Z})$ is *weakly stationary*, if for all $t, s \in \mathbb{Z}$

  (1) $\mathbf{E}x'_t x_t < \infty$
  (2) $\mathbf{E}x_t = \mathbf{E}x_0$
  (3) $\mathbf{E}x_t x'_s = \mathbf{E}x_{t-s} x'_0$ (or equivalently $(\mathbf{Cov}(x_t, x_s) = \mathbf{Cov}(x_{t-s}, x_0))$).

This book deals almost exclusively with weakly stationary processes. Thus, "stationary" will always mean "weak-sense stationary" in the following.

Readers should convince themselves that the assumption $\mathbf{E}x'_t x_t < \infty, \forall t \in \mathbb{Z}$ guarantees the existence of all first and second moments of the process. As is easy to see, neither strict stationarity implies weak stationarity, nor vice versa. The significance of the concept of weak stationarity results from the fact that the second moments of the process contain essential information about the underlying process. In particular, linear least squares approximations (as in prediction and filtering) can be determined from the knowledge of these second moments alone.

Stationarity means that essential properties of the process, the finite-dimensional marginal distributions in the case of strict stationarity or the first and second moments in the case of weak stationarity, respectively, are *invariant* to temporal translations. Stationary processes can be observed in stable random systems fed by constant energy. This is the case, for example, with ocean waves or machine vibrations. Many phenomena such as human speech or EEG signals can be described locally by stationary processes, even if they show clear non-stationarities. Non-stationary phenomena can often be traced back to stationary ones by transformations such as first differences or trend adjustment.

The invariance to temporal translations makes it possible, for example, to draw conclusions from the past about the future. Without assumptions such as stationarity, a meaningful statistical analysis would not be possible in many cases.

Definition 1.4 The *(auto)covariance function* of a (weakly) stationary process $(x_t \mid t \in \mathbb{Z})$ is the function

$$\begin{aligned} \gamma : \mathbb{Z} &\longrightarrow \mathbb{R}^{n \times n} \\ k &\longmapsto \gamma(k) = \mathbf{Cov}(x_{t+k}, x_t) \end{aligned} \tag{1.1}$$

and the *(auto)correlation function (ACF)* is

$$\begin{aligned} \rho : \mathbb{Z} &\longrightarrow \mathbb{R}^{n \times n} \\ k &\longmapsto \rho(k) = \mathbf{Corr}(x_{t+k}, x_t). \end{aligned} \tag{1.2}$$

The (i, j)th element of $\rho(k)$ is the correlation of $x_{i,t+k}$ and $x_{j,t}$, i.e.

$$\rho_{ij}(k) = \mathbf{Corr}(x_{i,t+k}, x_{j,t}) = \frac{\gamma_{ij}(k)}{\sqrt{\gamma_{ii}(0)\gamma_{jj}(0)}}$$

(and hence $\rho_{ii}(k) = \frac{\gamma_{ii}(k)}{\gamma_{ii}(0)}$). The correlations are bounded, $|\rho_{ij}(k)| \le 1$, and of course $\rho_{ii}(0) = 1$. The following symmetry property of the covariance function applies:

$$\gamma(k) = \mathbf{Cov}(x_k, x_0) = \mathbf{E}x_k x_0' - \mathbf{E}x_k \mathbf{E}x_0' = \mathbf{Cov}(x_0, x_k)' = \gamma(-k)'.$$

We use the same symbol γ for the covariance function of general processes and for the covariance function of stationary processes, which only depends on the lag $k = t - s$.

The covariance function or the correlation function respectively describe the dependencies between all pairs x_{it} and x_{js} and is the focus of the analysis and theory of (weakly) stationary processes.

We often consider "stacked" random vectors of the form

$$x_t^k := (x_t', x_{t-1}', \dots, x_{t+1-k}')'. \tag{1.3}$$

The covariance matrix of these random vectors is

$$\begin{aligned}
\Gamma_k :&= \mathbf{Var}(x_{t-1}^k) = \big(\mathbf{Cov}(x_{t-i}, x_{t-j})\big)_{i,j=1,\dots,k} = (\gamma(j-i))_{i,j=1,\dots,k} \\
&= \begin{pmatrix}
\gamma(0) & \gamma(1) & \gamma(2) & \cdots & \gamma(k-1) \\
\gamma(-1) & \gamma(0) & \gamma(1) & \cdots & \gamma(k-2) \\
\gamma(-2) & \gamma(-1) & \gamma(0) & \cdots & \gamma(k-3) \\
\vdots & \vdots & \vdots & \ddots & \vdots \\
\gamma(1-k) & \gamma(2-k) & \gamma(3-k) & \cdots & \gamma(0)
\end{pmatrix} \in \mathbb{R}^{nk \times nk}.
\end{aligned} \tag{1.4}$$

These matrices are covariance matrices and thus are always positive semidefinite and symmetric. Stationarity is reflected in the block Toeplitz matrix structure of the matrices, i.e. their (i, j)-block depends only on $(j - i)$.

Definition 1.5 A function $a \colon \mathbb{Z} \longrightarrow \mathbb{R}^{n \times n}$ is said to be *positive semidefinite*, if the matrices

$$A_k = \begin{pmatrix}
a(0) & a(1) & a(2) & \cdots & a(k-1) \\
a(-1) & a(0) & a(1) & \cdots & a(k-2) \\
a(-2) & a(-1) & a(0) & \cdots & a(k-3) \\
\vdots & \vdots & \vdots & \ddots & \vdots \\
a(1-k) & a(2-k) & a(3-k) & \cdots & a(0)
\end{pmatrix} \in \mathbb{R}^{nk \times nk}$$

are symmetric and positive semidefinite for each $k \in \mathbb{N}$.

Of course, the condition $A_k = A_k'$ implies, in particular, $a(k) = a(-k)'$. The following theorem characterizes covariance functions:

Proposition 1.6 *A function $a \colon \mathbb{Z} \longrightarrow \mathbb{R}^{n \times n}$ is the covariance function of a stationary process if and only if it is positive semidefinite.*

Proof From the above, it can be seen that the covariance function of a stationary process is positive semidefinite. The opposite direction can be seen here as follows: The positive semidefinite symmetric matrices A_k define nk-dimensional normal distributions and these form a consistent system of finite-dimensional normal distributions for varying k. According to Kolmogorov's consistency theorem,[5] a (Gaussian) process exists whose covariance function is a. See for example (Schmidt 2009, Folgerung 20.2.2).   □

If (y_t) and (z_t) are two stationary processes, then the "stacked" process $x_t = (y_t', z_t')'$ is stationary if and only if (y_t) and (z_t) are stationarily correlated to each other, i.e. if

$$\mathbf{Cov}(y_{t+k}, z_t) = \mathbf{Cov}(y_{s+k}, z_s) \; \forall t, s, k \in \mathbb{Z}.$$

In this case, the covariance function $\gamma_x(\cdot)$ of (x_t) can be partitioned accordingly as

$$\gamma_x(k) = \begin{pmatrix} \gamma_y(k) & \gamma_{yz}(k) \\ \gamma_{zy}(k) & \gamma_z(k). \end{pmatrix}$$

The diagonal blocks $\gamma_y(\cdot)$ and $\gamma_z(\cdot)$ are the (auto)covariance functions of the processes (y_t) or (z_t), respectively, and $\gamma_{yz}(\cdot)$ is the so-called *cross-covariance function* between the processes (y_t) and (z_t). One can partition (and interpret) the correlation function in the same way (Fig. 1.1).

Exercise 1.7 Consider the process $(x_t = \cos(\lambda t) \mid t \in \mathbb{N})$, where λ is a random variable uniformly distributed on the interval $[-\pi, \pi]$. For the sake of simplicity, we consider the process only on $\mathbb{N}$.

(1) Outline a few "typical trajectories" of the process.
(2) Calculate the mean value function $\mathbf{E}x_t$ and the autocovariance function $\gamma(t, s) = \mathbf{Cov}(x_t, x_s)$. Is the process weakly stationary? Note:

$$\cos(a)\cos(b) = \frac{1}{2}(\cos(a+b) + \cos(a-b)).$$

---

[5] Andrey N. Kolmogorov (1903–1987). Russian mathematician. Kolmogorov is considered to be one of the most outstanding mathematicians of the twentieth century. His most well-known achievement is the axiomatization of probability theory. Significant contributions to the theory of stationary processes.

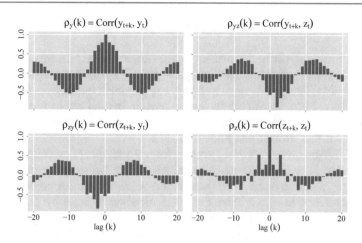

Fig. 1.1 Correlation function of a bivariate process ($x_t = (y_t, z_t)'$). The figures illustrate the autocorrelation function of (y_t) (top-left), the cross-correlation function between (y_t) and (z_t) (top-right), the cross-correlation function between (z_t) and (y_t) (bottom-left) and the autocorrelation function of (z_t) (bottom-right)

(3) Is the process strictly stationary? Note: Draw the curve(s) $\omega \in [-\pi, \pi] \mapsto (\cos(\omega t), \cos(\omega s)) \in \mathbb{R}^2$ for different $t, s \in \mathbb{N}$.

Exercise 1.8 Let A and B be two real-valued random variables. We now define the process ($x_t = A + (-1)^t B \mid t \in \mathbb{Z}$). Calculate the expectations $\mathbf{E}x_t$ and the covariance function $\gamma(t, s) = \mathbf{Cov}(x_t, x_s)$. Which conditions must be satisfied by A and B, so that these expectations and the covariances exist? Under which conditions is the process weakly stationary? Draw some "typical" trajectories of the process.

Exercise 1.9 Let there be an n-dimensional, weakly stationary process ($x_t \mid t \in \mathbb{Z}$) with the expected value $\mu_x = \mathbf{E}x_t$ and the covariance functions $\gamma_x(k)$. Show that the process ($y_t \mid t \in \mathbb{Z}$), which is defined by $y_t = c + b_0 x_t + b_1 x_{t-1}$ for $c \in \mathbb{R}^m$ and $b_0, b_1 \in \mathbb{R}^{m \times n}$, is also weakly stationary and calculate the expectation $\mathbf{E}y_t$ and the covariance function $\gamma_y(k) = \mathbf{Cov}(y_{t+k}, y_t)$.

Exercise 1.10 Let there be a scalar Gaussian process ($x_t \mid t \in \mathbb{Z}$), i.e. a process for which $(x_{t_1}, \ldots, x_{t_s})'$ for all finite subsets $\{t_1, \ldots, t_s\} \subset \mathbb{Z}, s > 0$ is multivariate normally distributed. Show that (x_t) is strictly stationary if and only if (x_t) is weakly stationary. Now let us suppose that (x_t) is (weakly and strictly) stationary with the expected value $\mu_x = \mathbf{E}x_t$ and the covariance function $\gamma_x(k) = \mathbf{Cov}(x_{t+k}, x_t)$. Show that the process ($y_t = \exp(x_t) \mid t \in \mathbb{Z}$) is weakly stationary and calculate the expectation $\mu_y = \mathbf{E}y_t$ and the covariance function $\gamma_y(k) = \mathbf{Cov}(y_{t+k}, y_t)$. Note: $\log(y_t y_s)$ is normally distributed. $\mathbf{E}\left[(\exp(u))^k\right] = \exp(k\mu + \frac{k^2 \sigma^2}{2})$ holds for a normally distributed random variable $u \sim N(\mu, \sigma^2)$.

1.3   The Time Domain of Stationary Processes

Stationary processes can be "embedded" in the Hilbert space of the square integrable random variables over the probability space $(\Omega, \mathcal{A}, \mathbf{P})$. This makes it possible to give (geometric) interpretations as well as to use important results from the theory of Hilbert spaces. First, we will briefly review some terms and results from the theory of Hilbert spaces and, in particular, of the Hilbert space of square integrable random variables (see, e.g. Brokate and Kersting 2011):

A *real Hilbert space* is a linear space with real multipliers, which has an inner product and which is complete with respect to the metric defined by the inner product. Later on, we will also discuss complex Hilbert spaces. The inner product between two vectors $x, y \in \mathbb{H}$ of a Hilbert space $\mathbb{H}$ is denoted as usual by $\langle x, y \rangle$ and the corresponding norm by $\|x\| = \sqrt{\langle x, x \rangle}$. Two vectors $x, y \in \mathbb{H}$ are orthogonal, if $\langle x, y \rangle = 0$ and we write $x \perp y$. Likewise, we say $x \in \mathbb{H}$ is orthogonal to a subset $\mathbb{M} \subset \mathbb{H}$ if $x \perp y \ \forall y \in \mathbb{M}$ and we use the notation $x \perp \mathbb{M}$. Two subsets $\mathbb{M}_1 \subset \mathbb{H}$ and $\mathbb{M}_2 \subset \mathbb{H}$ are called orthogonal (with symbols $\mathbb{M}_1 \perp \mathbb{M}_2$), if $x_1 \perp x_2$ for all $x_1 \in \mathbb{M}_1$ and $x_2 \in \mathbb{M}_2$.

A central result, which we will use repeatedly, is the projection theorem (see, e.g. Brokate and Kersting 2011):

Proposition 1.11 (projection theorem) Let $\mathbb{H}$ be a *Hilbert space* and $\mathbb{M} \subset \mathbb{H}$ *a closed subspace of* $\mathbb{H}$. *Then*

(1) Each $x \in \mathbb{H}$ can be represented in exactly one way as

$$x = y + z$$

where $y \in \mathbb{M}$ and $z \perp \mathbb{M}$.
(2) The element y is (in terms of the metric as defined by the inner product) the best approximation of x in $\mathbb{M}$, i.e.

$$\|x - y\| = \min_{\tilde{y} \in \mathbb{M}} \|x - \tilde{y}\|.$$

The element y is called *projection* of x onto $\mathbb{M}$, and z is the corresponding *perpendicular*. For the perpendicular, we therefore have

$$\|z\|^2 = \|x - y\|^2 = \|x\|^2 - \|y\|^2.$$

The map, which attaches the projection y to $x \in \mathbb{H}$, is called projector on $\mathbb{M}$ and is often denoted by $P_\mathbb{M}$. One can easily see that $P_\mathbb{M}$ is a linear map.

If $\mathbb{M}_1 \subset \mathbb{H}$ and $\mathbb{M}_2 \subset \mathbb{H}$ are two closed subspaces of a Hilbert space that are *orthogonal* to each other ($\mathbb{M}_1 \perp \mathbb{M}_2$) and if $\mathbb{M} = \mathbb{M}_1 \oplus \mathbb{M}_2$ denotes the direct sum of $\mathbb{M}_1$ and $\mathbb{M}_2$, then

$$P_\mathbb{M} = P_{\mathbb{M}_1} + P_{\mathbb{M}_2}.$$

A concrete Hilbert space is of particular importance for us. Let $(\Omega, \mathcal{A}, \mathbf{P})$ be a probability space. In the following, we shall also use the term random variable for the equivalence class of the $\mathbf{P}$-almost surely identical measurable functions $x : \Omega \longrightarrow \mathbb{R}$. One can prove that the set of all (equivalence classes of) square integrable (one-dimensional) random variables (i.e. $\mathbf{E}x^2 < \infty$) with the usual addition and scalar multiplication and with the inner product

$$\langle x, y \rangle = \mathbf{E}xy = \int xy \, d\mathbf{P}$$

forms a Hilbert space, which is usually denoted by $\mathbb{L}_2(\Omega, \mathcal{A}, \mathbf{P})$ or in brief by $\mathbb{L}_2$. The completeness of this space is shown in the so-called Riesz–Fischer theorem (see, e.g. Brokate and Kersting 2011, Satz VI.2).

The inner product of this Hilbert space defines the norm

$$\|x\| = \sqrt{\mathbf{E}x^2} \tag{1.5}$$

and the corresponding convergence, the so-called *convergence in mean square:* A sequence $(x_k \in \mathbb{L}_2)_{k \geq 1}$ converges in mean square to the limiting function $x_0 \in \mathbb{L}_2$, if

$$\lim_{k \to \infty} \|x_k - x_0\|^2 = \lim_{k \to \infty} \mathbf{E}(x_k - x_0)^2 = 0.$$

We use the notation $x_0 = \text{l.i.m}_{k \to \infty} x_k$. Due to the completeness of $\mathbb{L}_2$, the Cauchy criterion applies to this convergence:

$$(x_k \in \mathbb{L}_2)_{k \geq 1} \text{ converges in mean square} \iff \lim_{k,l \to \infty} \mathbf{E}(x_k - x_l)^2 = 0.$$

Suppose $(x_k \in \mathbb{L}_2)_{k \geq 1}$ and $(y_k \in \mathbb{L}_2)_{k \geq 1}$ are two convergent sequences with $x_0 = \text{l.i.m}_k x_k$ and $y_0 = \text{l.i.m}_k y_k$. Then the continuity of the inner product implies

$$\lim_{k \to \infty} \mathbf{E}x_k y_k = \mathbf{E}\left[(\text{l.i.m}_{k \to \infty} x_k)(\text{l.i.m}_{k \to \infty} y_k) \right] = \mathbf{E}x_0 y_0$$

and

$$\lim_{k \to \infty} \mathbf{E}x_k = \lim_{k \to \infty} \mathbf{E}[1 x_k] = \mathbf{E}\left[1 \, \text{l.i.m}_{k \to \infty} x_k \right] = \mathbf{E}x_0.$$

Here and in the following, $1 \in \mathbb{L}_2$ denotes the random variable, which is equal to $1 \in \mathbb{R}$ a.s. This means that expectations, variances and covariances are continuous with respect to convergence in the mean square.

Many terms and concepts in statistics have an equivalent in the Hilbert space $\mathbb{L}_2$ and hence have a "geometrical" interpretation. This "translation" is particularly simple for the case of *centered* random variables (i.e. random variables with expectations equal to zero) :

- The norm of x is equal to the standard deviation: $\|x\| = \sqrt{\mathbf{Var}(x)}$.
- The inner product is equal to the covariance: $\langle x, y \rangle = \mathbf{Cov}(x, y)$.
- The correlation is the cosine of the "angle between the two random variables":

$$\mathbf{Corr}(x, y) = \frac{\mathbf{Cov}(x, y)}{\sqrt{\mathbf{Var}(x)\mathbf{Var}(y)}} = \frac{\langle x, y \rangle}{\|x\|\|y\|} = \cos(\measuredangle(x, y)).$$

- Uncorrelated means orthogonal in $\mathbb{L}_2$: $\mathbf{Cov}(x, y) = 0$ is equivalent to $x \perp y$.
- The random variables $x_1, \ldots, x_k \in \mathbb{L}_2$ are linearly independent in $\mathbb{L}_2$, if and only if the covariance matrix $\mathbf{Var}((x_1, \ldots, x_k)')$ is positive definite.

Exercise 1.12 (*Exercises relating to* $\mathbb{L}_2$) Show the following:

(1) Suppose $x_1, \ldots, x_k \in \mathbb{L}_2$ are centered random variables ($\mathbf{E}x_i = 0$). The rank of the covariance matrix $\Gamma = \mathbf{Var}((x_1, \ldots, x_k)') = \mathbf{E}(x_1, \ldots, x_k)(x_1, \ldots, x_k)'$ equals the dimension of the span $\mathrm{sp}\{x_1, \ldots, x_k\} \subset \mathbb{L}_2$.
(2) Projectors are linear, idempotent and self-adjoint mappings. The eigenvalues of projectors are 0 and 1.
(3) If $\mathbb{M}_1$, $\mathbb{M}_2 \subset \mathbb{L}_2$ are two closed subspaces of a Hilbert space that are orthogonal to each other, then $P_{\mathbb{M}_1 \oplus \mathbb{M}_2} = P_{\mathbb{M}_1} + P_{\mathbb{M}_2}$.

In the same way, random variables with an expectation not equal to zero can be treated. In particular, $\mathbf{E}x = \mathbf{E}1x = \langle 1, x \rangle$ and $\mathbf{E}x = 0$ is equivalent to $x \perp 1$.

The Cauchy–Schwarz inequality in $\mathbb{L}_2$ is

$$|\mathbf{E}xy| = |\langle x, y \rangle| \le \|x\|\|y\| = \sqrt{\mathbf{E}x^2}\sqrt{\mathbf{E}y^2}$$

and the Pythagorean theorem is

$$\mathbf{E}(x + y)^2 = \|x + y\|^2 = \|x\|^2 + \|y\|^2 = \mathbf{E}x^2 + \mathbf{E}y^2 \text{ if } \mathbf{E}xy = 0.$$

If $(z_k \in \mathbb{L}_2)_{k \ge 1}$ is a generating set of a closed subspace of a Hilbert space $\mathbb{M} \subset \mathbb{L}_2$, then the projection $y = P_{\mathbb{M}} x$ of $x \in \mathbb{L}_2$ onto $\mathbb{M}$ is referred to as the best linear least squares approximation of x by $(z_k)_{k \ge 1}$. It is linear because each element of $\mathbb{M}$ can be represented as a linear combination of z_k's or as the limit of such linear combinations. Moreover, if $(z_k \in \mathbb{L}_2)_{k \ge 1}$ is an orthonormal basis of $\mathbb{M}$, then the projection y of x onto $\mathbb{M}$ can be represented as

$$y = \sum_{k=1}^{\infty} (\mathbf{E}xz_k) z_k$$

and Parseval's identity holds

$$\|y\|^2 = \sum_{k=1}^{\infty} (\mathbf{E}xz_k)^2.$$

The best linear least squares approximation of x by an element y from $\mathbb{M}$ is (fully) characterized by

$$y \in \mathbb{M} \text{ and } \mathbf{E}(x - y)z_k = 0 \ \ k = 1, 2, \ldots$$

This characterization holds for any (not necessarily orthonormal) generating set.

Exercise 1.13 (*projection*) Suppose $x, z_1, \ldots, z_n \in \mathbb{L}_2$ and $\mathbb{M} = \mathrm{sp}(z_1, \ldots, z_n)$, $\Sigma_{zz} = \mathbf{E}(z_1, \ldots, z_n)(z_1, \ldots, z_n)' \in \mathbb{R}^{n \times n}$, $\Sigma_{xz} = \mathbf{E}x(z_1, \ldots, z_n) \in \mathbb{R}^{1 \times n}$ and $c \in \mathbb{R}^n$. Show that

(1) $y = c'(z_1, \ldots, z_n)'$ is the projection of x onto $\mathbb{M}$, if and only if

$$c'\Sigma_{zz} = \Sigma_{xz}.$$

For the corresponding perpendicular, the following holds:

$$\mathbf{E}(x - y)^2 = \|x - y\|^2 = \|x\|^2 - \|y\|^2 = \mathbf{E}x^2 - c'\Sigma_{zz}c.$$

(2) The above system of equations can always be solved for c. It can be solved uniquely if $\Sigma_{zz} > 0$, i.e. if $\{z_1, \ldots, z_k\}$ is a basis for $\mathbb{M}$. In this case, $c' = \Sigma_{xz}\Sigma_{zz}^{-1}$ and $\mathbf{E}(x - y)^2 = \mathbf{E}x^2 - \Sigma_{xz}\Sigma_{zz}^{-1}\Sigma_{xz}'$.

(3) Provided that Σ_{zz} is singular, then, e.g. $c' = \Sigma_{xz}\Sigma_{zz}^{\dagger}$ is a solution, where $\Sigma_{zz}^{\dagger}$ denotes the Moore–Penrose inverse of Σ_{zz}. The projection y and the corresponding perpendicular $(x - y)$ are unique also when $\det \Sigma_{zz} = 0$ holds.

After this brief recap, back to the stationary process (x_t), which will be embedded in the Hilbert space $\mathbb{L}_2$ as follows:

Suppose $x_t = (x_{1t}, \ldots, x_{nt})'$. From (1) in the definition of weak stationarity, we get $x_{it} \in \mathbb{L}_2$, $i = 1, \ldots, n$, $t \in \mathbb{Z}$. Condition (3) implies that all elements x_{it} of the ith sub-process $(x_{it} \mid t \in \mathbb{Z})$ are of equal length $\|x_{it}\|$, the ith sub-process thus "runs" on a sphere in Hilbert space $\mathbb{L}_2$. Moreover, the angles between x_{it} and x_{js} depend only on $t - s$. Condition (2) implies that the angles between x_{it} and 1 do not depend on t.

Definition 1.14 (*time domain*) Suppose $(x_t \mid t \in \mathbb{Z})$ is a stationary process, then the closed subspace $\mathbb{H}(x)$ generated by $\{x_{it} \mid i = 1, \ldots, n, \ t \in \mathbb{Z}\}$ in $\mathbb{L}_2$ is the called the *time domain* of the process (x_t).

By definition, the time domain is the closed subspace of $\mathbb{L}_2$, which consists of all linear combinations

$$\sum_{j=-N}^{N} a_j' x_{t-j}, \ a_j \in \mathbb{R}^n$$

and their limits. In many cases, but not always, such limits can be represented as infinite sums (in the sense of convergence in mean square)

$$\sum_{j=-\infty}^{\infty} a'_j x_{t-j}.$$

We have $\mathbb{H}(x) = \overline{\text{sp}}\{x_{it} \mid i = 1, \ldots, n, \ t \in \mathbb{Z}\}$, where $\overline{\text{sp}}\{\}$ denotes the closure of the span $\text{sp}\{\}$ of the respective generating set and thus is a Hilbert space of its own. The time domain $\mathbb{H}(x)$ is the smallest Hilbert subspace of $\mathbb{L}_2$, which contains all one-dimensional process variables x_{it}.

The Hilbert space "geometry" of stationary processes as described above suggests that the mapping

$$x_{it} \longmapsto x_{i,t+1}, \ i = 1, \ldots, n, \ t \in \mathbb{Z}$$

can be extended to a unitary operator on $\mathbb{H}(x)$. An operator

$$\text{U} \colon \mathbb{H} \longrightarrow \mathbb{H}$$

(on a Hilbert space $\mathbb{H}$) is called *unitary* if U *is bijective* and *isometric*. The latter means that

$$\langle \text{U} x, \text{U} y \rangle = \langle x, y \rangle \ \forall x, y \in \mathbb{H}.$$

Every unitary operator is linear and continuous as can be easily shown.

Proposition 1.15 (forward shift) *The operator defined by*

$$x_{it} \longmapsto \text{U} x_{it} = x_{i,t+1}, \ i = 1, \ldots, n, \ t \in \mathbb{Z}$$

can be uniquely extended to a unitary operator on $\mathbb{H}(x)$. *This operator is called* forward shift *of the process* (x_t).

Proof Due to the stationarity of (x_t), $\langle \text{U} x_{it}, \text{U} x_{js} \rangle = \langle x_{i,t+1}, x_{j,s+1} \rangle = \langle x_{it}, x_{js} \rangle$ holds for all $i, j = 1, \ldots, n$ and $t, s \in \mathbb{Z}$. U is also well-defined on $\{x_{it} \mid i = 1, \ldots, n, \ t \in \mathbb{Z}\}$ (i.e. U is a function), because $x_{it} = x_{js}$ implies $\mathbf{E}(x_{it} - x_{js})^2 = 0$ and thus by stationarity $\mathbf{E}(x_{i,t+1} - x_{j,s+1})^2 = 0$, hence $\text{U} x_{it} = x_{i,t+1} = x_{j,s+1} = \text{U} x_{js}$. As easily can be seen the linear extension

$$\text{U}\left(\sum_{k=1}^{m} a_k x_{i_k, t_k}\right) = \sum_{k=1}^{m} a_k \, \text{U} \, x_{i_k, t_k} = \sum_{k=1}^{m} a_k x_{i_k, t_k+1}$$

to the linear span $\text{sp}\{x_{it} \mid i = 1, \ldots, n, \ t \in \mathbb{Z}\}$ is also well-defined and isometric. For any $y \in \mathbb{H}(x)$ there exists a sequence $y^{(m)} \in \text{sp}\{x_{it} \mid i = 1, \ldots, n, \ t \in \mathbb{Z}\}$ such that $y^{(m)} \to y$ and we define the continuous extension of U on $\mathbb{H}(x)$ by $\text{U} y = \text{l.i.m}_m \, \text{U} \, y^{(m)}$. Again, we can easily see that this mapping is well-defined and isometric on $\mathbb{H}(x)$. Similarly, the backward shift $U^{-1} : \mathbb{H}(x) \longrightarrow \mathbb{H}(x)$ can be constructed as an (unitary) extension of the mapping $x_{it} \longmapsto x_{i,t-1}$. The backward shift is the inverse function of the forward shift and both functions are hence bijective.   □

We denote the composition of the forward and the backward shift, respectively, by (U^0), which means that (U^0) is the identity on $(\mathbb{H}(x))$. Furthermore let $(U^t = U U^{t-1})$ for $(t > 0)$ and $(U^t = U^{-1} U^{t+1})$ for $(t < 0)$. Of course, $U^{t+s} = U^t U^s$ holds for $t, s \in \mathbb{Z}$.

We can apply the forward shift U and its powers componentwise to random vectors $y = (y_1, \dots, y_m)'$, $y_i \in \mathbb{H}(x)$, that is

$$U^t y := (U^t y_1, \dots, U^t y_m)'.$$

Clearly,

$$x_t = U^t x_0, \ \forall t \in \mathbb{Z}$$

holds. Thus, the process (x_t) is obtained from the "initial value" x_0 and the iterative application of the forward and backward shift, respectively, on this initial value. In addition for every random vector $y_0 = (y_{10}, \dots, y_{m0})'$, $y_{i0} \in \mathbb{H}(x)$

$$\left(y_t = U^t y_0 \mid t \in \mathbb{Z} \right)$$

defines a stationary process that is stationary correlated to the original process, i.e. the process $(z_t = (x_t', y_t')')$ is stationary.

Exercise 1.16 Suppose $(x_t = (x_{1t}', x_{2t}')' \mid t \in \mathbb{Z})$ is a stationary process and U is the forward shift of (x_t). Show that the restriction of U to $\mathbb{H}(x_1) = \overline{\mathrm{sp}}\{x_{1s} \mid s \in \mathbb{Z}\} \subset \mathbb{H}(x)$ is the forward shift of the (sub)process (x_{1t}).

In the following, we will often deal with random vectors whose components are elements of $\mathbb{L}_2$, in particular, of course, the random vectors x_t corresponding to a multivariate stationary process (x_t). Hence, we introduce the following conventions: Let $\mathbb{M} \subset \mathbb{L}_2$ be a subset of $\mathbb{L}_2$. $\mathbb{M}^p$ denotes the set of p-tuples with elements in $\mathbb{M}$. That is, $u \in \mathbb{M}^p$ means $u = (u_1, \dots, u_p)'$ and $u_i \in \mathbb{M}$. Therefore, $\mathbb{L}_2^p$, in particular, is the set of random vectors of dimension p with components in $\mathbb{L}_2$.

For $u \in \mathbb{L}_2^p$, $v \in \mathbb{L}_2^q$, we now define the following:

- $\langle u, v \rangle := \mathbf{E}uv' \in \mathbb{R}^{p \times q}$.
- $u \perp v$ means $u_i \perp v_j$ for all $i = 1, \dots, p$ and $j = 1, \dots, q$. This condition is equivalent to $\langle u, v \rangle = \mathbf{E}uv' = 0 \in \mathbb{R}^{p \times q}$.
- $u \perp \mathbb{M} \subset \mathbb{L}_2$ means $u_i \perp \mathbb{M}$ for $i = 1, \dots, p$.
- The projection of $u \in \mathbb{L}_2^p$ onto a closed subspace of a Hilbert space $\mathbb{M} \subset \mathbb{L}_2$ is defined componentwise

$$P_{\mathbb{M}} u = \begin{pmatrix} P_{\mathbb{M}} u_1 \\ \vdots \\ P_{\mathbb{M}} u_p \end{pmatrix}. \tag{1.6}$$

Due to the linearity of the projection, for every matrix $A \in \mathbb{R}^{q \times p}$, we have

$$\mathrm{P}_{\mathbb{M}}(Au) = A(\mathrm{P}_{\mathbb{M}} u).$$

- Suppose $(w_k = (w_{1k}, \dots, w_{pk})' \in \mathbb{L}_2^p \mid k \in I \subset \mathbb{Z})$: For subspaces of the form $\overline{\mathrm{sp}}\{w_{ik} \mid i = 1, \dots, p, \ k \in I\}$, we simply write $\overline{\mathrm{sp}}\{w_k \mid k \in I \subset \mathbb{Z}\}$, e.g. $\mathbb{H}(x) = \overline{\mathrm{sp}}\{x_t \mid t \in \mathbb{Z}\}$.
- Convergence in mean square of sequences $w_k \in \mathbb{L}_2^p$ to $w_0 \in \mathbb{L}_2^p$ is defined (componentwise) by

$$w_0 = \underset{k\to\infty}{\mathrm{l.i.m}}\, w_k \iff w_{i0} = \underset{k\to\infty}{\mathrm{l.i.m}}\, w_{ik} \iff \lim_{k\to\infty} \mathbf{E}(w_k - w_0)'(w_k - w_0) = 0.$$

A natural question is why one does not provide $\mathbb{L}_2^n$ with a suitable Hilbert space structure and then interpret vectors x_t as elements of this Hilbert space. Here, the problem is that for linear least squares approximations such as the approximation of $x_{t+1} \in \mathbb{L}_2^n$ by a linear combination of the form $\sum_{j=1}^k a_j x_{t+1-j}$, the square $n \times n$ matrices ($a_j \in \mathbb{R}^{n \times n}$) are natural multipliers. However, these *do not* form a field. This means that $\mathbb{L}_2^n$, with matrices as multipliers, is not a linear space and thus not a Hilbert space either. On the other hand, the restriction to scalar multipliers $a_j \in \mathbb{R}$ would be a too strong (and unnecessary) restriction in many cases.

1.4   Examples of Stationary Processes

We will discuss important classes of stationary processes in this section, including white noise, MA processes, AR processes as well as harmonic processes.

Definition 1.17 (*white noise*) An (n-dimensional) process ($\epsilon_t \mid t \in \mathbb{Z}$) is called *white noise*, if for all $t, s \in \mathbb{Z}$

(1) $\mathbf{E}\epsilon_t'\epsilon_t < \infty$.
(2) $\mathbf{E}\epsilon_t = 0 \in \mathbb{R}^n$.
(3) $\mathbf{E}\epsilon_t\epsilon_t' = \mathbf{E}\epsilon_0\epsilon_0' = \Sigma \in \mathbb{R}^{n \times n}$.
(4) $\mathbf{E}\epsilon_t\epsilon_s' = 0 \in \mathbb{R}^{n \times n}$ for $t \neq s$.

We often use the notation $(\epsilon_t) \sim \mathrm{WN}(\Sigma)$ for white noise with variance $\mathbf{E}\epsilon_t\epsilon_t' = \Sigma$. Clearly, white noise is weakly stationary. White noise has no linear dependencies over time (no (linear) memory), since $\mathbf{E}\epsilon_{t+k}\epsilon_t' = 0$ for $k \neq 0$. Hence, they do not play a major role for modeling practically relevant phenomena. As we will see very soon, they are mainly used as "building blocks" for more complex processes.

Moving Average Processes

Definition 1.18 *(MA(q) process)* Let $(\epsilon_t) \sim \mathrm{WN}(\Sigma)$ be m-dimensional, white noise and $b_0, b_1, \ldots, b_q \in \mathbb{R}^{n \times m}$ with $b_0 \Sigma b_q' \neq 0$. In this case, the process defined by

$$x_t = b_0 \epsilon_t + \cdots + b_q \epsilon_{t-q}, \quad t \in \mathbb{Z} \tag{1.7}$$

is called *moving average* process of the order q *(in short: MA(q)* process).

An MA(q) process is weakly stationary with the mean value function

$$\mathbf{E} x_t = \mathbf{E}(b_0 \epsilon_t + \cdots + b_q \epsilon_{t-q}) = 0$$

and covariance function

$$
\begin{aligned}
\gamma(k) &= \mathbf{Cov}(x_{t+k}, x_t) \\
&= \mathbf{E}(b_0 \epsilon_{t+k} + \cdots + b_q \epsilon_{t+k-q})(b_0 \epsilon_t + \cdots + b_q \epsilon_{t-q})' \\
&= \sum_{i,j=0}^{q} b_i \mathbf{E}\left[\epsilon_{t+k-i} \epsilon_{t-j}'\right] b_j' \\
&= \begin{cases} \sum_{j=0}^{q-k} b_{j+k} \Sigma b_j' & \text{for } 0 \leq k \leq q \\ \sum_{j=-k}^{q} b_{j+k} \Sigma b_j' & \text{for } -q \leq k < 0 \\ 0 & \text{for } |k| > q. \end{cases}
\end{aligned}
$$

In the double sum in the third row above, all terms are zero except for those for which $t + k - i = t - j$ holds, that is $i = j + k$. If we define $b_j = 0 \in \mathbb{R}^{n \times m}$ for $j < 0$ and $j > q$, we obtain a common formula for the covariances for all lags k

$$\gamma(k) = \sum_{j=0}^{q} b_{j+k} \Sigma b_j'. \tag{1.8}$$

MA(q) processes thus have a "finite linear memory", since $\gamma(k) = 0$ for $|k| > q$. Conversely, one can also show that a stationary process with $\gamma(q) \neq 0$ and $\gamma(k) = 0 \ \forall |k| > q$ is an MA(q) process, which means that it has a representation of the form (1.7), where w.l.o.g one may additionally assume $m = n$ and $b_0 = I_n$. See exercise 2.15 on page 42.

Exercise 1.19 In the following, we will also consider "two-sided" MA processes of the form $x_t = \sum_{j=-q}^{q} b_j \epsilon_{t-j}$. Show that (x_t) is stationary with $\mathbf{E} x_t = 0$ and $\gamma(k) = \mathbf{Cov}(x_{t+k}, x_t) = \sum_{j=-q}^{q-k} b_{j+k} \Sigma b_j'$ for $0 \leq k \leq 2q$, $\gamma(k) = 0$ for $k > 2q$ and $\gamma(k) = \gamma(-k)'$ for $k < 0$.

Exercise 1.20 (*Autocovariance function of a scalar MA(1) process*) Let $(x_t = \epsilon_t + b_1\epsilon_{t-1} \mid t \in \mathbb{Z})$ (with $(\epsilon_t) \sim WN(\Sigma)$ and $b_1 \in \mathbb{R}$) be a scalar MA(1) process with autocovariance function γ. Show (for $\Sigma > 0$)

$$\gamma(1)) \leq \frac{1}{2}\gamma(0).$$

Definition 1.21 (*MA(∞) process*) An MA(∞) process $(x_t \mid t \in \mathbb{Z})$ is a process which has a representation of the form

$$x_t = \sum_{j=-\infty}^{\infty} b_j\epsilon_{t-j} \tag{1.9}$$

where $(\epsilon_t) \sim WN(\Sigma)$ is m-dimensional white noise and the sequence $(b_j \in \mathbb{R}^{n \times m} \mid j \in \mathbb{Z})$ is square summable, i.e.

$$\sum_{j=-\infty}^{\infty} \|b_j\|^2 < \infty. \tag{1.10}$$

A process(x_t), which has a so-called *causal* representation of the form

$$x_t = \sum_{j \geq 0} b_j\epsilon_{t-j}, \tag{1.11}$$

is called *causal MA(∞)* process.

Here and in the following, $\|A\|$ denotes any matrix norm, which is equivalent to the Frobenius norm $\|A\|_F^2 = \mathrm{tr}(A'A)$, in the sense that there exist constants $c_1, c_2 > 0$ such that $c_1\|A\|_F \leq \|A\| \leq c_2\|A\|_F$ holds for any $A \in \mathbb{R}^{m \times n}$. (The constants c_1, c_2 may depend on the dimension (m, n).) In particular, we sometimes use the spectral norm $\|A\|_2^2 = \max_i\{\lambda_i(A'A)\}$, where $\lambda_i(A'A)$ denotes the ith eigenvalue of $A'A$. Note that the Frobenius and the spectral norm are both submultiplicative, which means that $\|AB\| \leq \|A\|\|B\|$ holds.

In principle, process and representation, such as an MA(∞) representation (1.9), have to be distinguished. This, in particular, is due to the fact that the representation is not unique without further restrictions. Note that not every stationary process has an MA(∞) representation, and not every MA(∞) process has a causal MA(∞) representation.

Proposition 1.22 *The infinite sum $\sum_{j=-\infty}^{\infty} b_j\epsilon_{t-j}$ exists (as a limit of the partial sums in $\mathbb{L}_2$) if and only if $\sum_{j=-\infty}^{\infty} \mathrm{tr}(b_j\Sigma b_j') < \infty$. For $\Sigma > 0$, this condition is equivalent to $\sum_{j=-\infty}^{\infty} \|b_j\|^2 < \infty$.*
MA(∞) processes are weakly stationary with

$$\mathbf{E}x_t = 0$$

and

$$\gamma(k) = \mathbf{Cov}(x_{t+k}, x_t) = \sum_{j=-\infty}^{\infty} b_{j+k}\Sigma b'_j. \tag{1.12}$$

Proof The partial sums $x_t^q := \sum_{j=-q}^q b_j \epsilon_{t-j}$ converge (in the $\mathbb{L}_2$-sense) if and only if they form a Cauchy sequence. This means that for every $\nu > 0$, there exists a $q \in \mathbb{N}$, such that for all $r \geq s \geq q$

$$\mathbf{E}(x_t^r - x_t^s)'(x_t^r - x_t^s) = \mathbf{E}(\sum_{s<|j|\leq r} b_j\epsilon_{t-j})'(\sum_{s<|j|\leq r} b_j\epsilon_{t-j}) = \sum_{s<|j|\leq r} \mathrm{tr}(b_j \Sigma b'_j) \leq \nu.$$

This, however, is exactly the condition for the convergence of the partial sums $\sum_{j=-q}^q \mathrm{tr}(b_j \Sigma b'_j)$. If Σ is positive definite, then two constants $c_1, c_2 > 0$ exist such that $c_2 I_n \geq \Sigma \geq c_1 I_n$ holds, and hence also $c_2 \|b_j\|_F^2 \geq \mathrm{tr}(b_j \Sigma b'_j) \geq c_1 \|b_j\|_F^2$. The condition $\sum_j \mathrm{tr}(b_j \Sigma b'_j) < \infty$ is therefore equivalent to $\sum_j \|b_j\|_F^2 < \infty$. We have used the following notation here: For symmetric matrices A, B, $A > B$ ($A \geq B$) means that $A - B$ is positive definite (resp. positive semidefinite).

We now suppose that coefficients b_j are square summable and hence the process (x_t) is well-defined. Of course, the random vector x_t is square integrable as a limit in mean square. The "partial sums process" $(x_t^q \mid t \in \mathbb{Z})$ is a "two-sided" MA process, as covered in the Exercise 1.19 above. Using the continuity of expectation and covariance with respect to the convergence in mean square, we have

$$\mathbf{E}x_t = \mathbf{E}\,\mathrm{l.i.m}_{q\to\infty} x_t^q = \lim_{q\to\infty} \mathbf{E}x_t^q = \lim_{q\to\infty} 0 = 0$$

and

$$\mathbf{Cov}(x_{t+k}, x_t) = \mathbf{Cov}\left(\mathrm{l.i.m}_{q\to\infty} x_{t+k}^q, \mathrm{l.i.m}_{q\to\infty} x_t^q\right)$$

$$= \lim_{q\to\infty} \mathbf{Cov}\left(x_{t+k}^q, x_t^q\right) = \lim_{q\to\infty} \sum_{j=\max(-q,-q-k)}^{\min(q,q-k)} b_{j+k}\Sigma b'_j$$

$$= \sum_{j=-\infty}^{\infty} b_{j+k}\Sigma b'_j.$$

Thus, the expectation $\mathbf{E}x_t$ and the covariances $\mathbf{Cov}(x_{t+k}, x_t)$ are independent of t and therefore the MA(∞) process is stationary. □

Exercise 1.23 Show that MA(∞) processes have "fading memory" meaning that $\gamma(k) \to 0$ for $|k| \to \infty$ holds. Hint: Decompose the process as $x_t = x_t^q + u_t^q$ and use the Cauchy–Schwarz inequality, which, for instance, for a scalar process (x_t) implies $|\langle x_{t+k}^q, u_t^q \rangle| \leq \|x_{t+k}^q\| \|u_t^q\|$.

MA(q) processes, like AR(p) and ARMA(p, q) processes discussed below, have the property that their second moments are described by finitely many parameters. This is of great advantage for statistical analysis. The class of MA(∞) processes is a very large one within the class of stationary processes, which includes, in particular, AR and ARMA processes. The Wold theorem, discussed in the next chapter, shows that every regular process has a causal MA(∞) representation. These regular processes play a dominant role in practice.

Autoregressive Processes

A linear difference equation system of the form

$$x_t = a_1 x_{t-1} + \cdots + a_p x_{t-p} + \epsilon_t, \ \forall t \in \mathbb{Z} \tag{1.13}$$

where $a_j \in \mathbb{R}^{n \times n}$ and $(\epsilon_t) \sim \text{WN}(\Sigma)$ is white noise is called *autoregressive system* (AR system). A stationary solution, i.e. a stationary process (x_t), which satisfies these equations for all $t \in \mathbb{Z}$, is called *autoregressive process* (AR process). The denotation "autoregressive" indicates that the value x_t of the process at time t is represented as a (linear) function of its own past and an error term. Certain intertemporal relationships are thus explicitly represented by this model. AR processes have a number of useful properties. In particular, prediction is very simple and estimation of such models is also relatively elementary.

However, the difference equation (1.13) is only an implicit description of the AR process. This raises the question as to whether a stationary solution exists and, if so, whether this stationary solution is unique. For a more detailed discussion of AR systems and AR processes, see Chap. 5. Here we consider only a simple special case, namely a scalar ($n = 1$) AR system with $p = 1$:

$$x_t = a x_{t-1} + \epsilon_t \tag{1.14}$$

where we suppose that $\sigma^2 = \mathbf{E}\epsilon_t^2 > 0$.

The case $a = 0$ is trivial, because $x_t = \epsilon_t$ holds in this case. This means that here, the difference equation (1.14) has exactly one solution and this solution is, of course, stationary.

If $a \neq 0$, one can easily obtain a solution by iterating from an "initial value" x_0, for $t > 0$:

$$x_1 = a x_0 + \epsilon_1$$
$$x_2 = a x_1 + \epsilon_2 = a^2 x_0 + \epsilon_2 + a\epsilon_1$$
$$\vdots$$
$$x_t = a^t x_0 + \sum_{j=0}^{t-1} a^j \epsilon_{t-j} \ \text{for } t > 0$$

and for $t < 0$:

$$x_{-1} = a^{-1}x_0 - a^{-1}\epsilon_0$$
$$x_{-2} = a^{-1}x_{-1} - a^{-1}\epsilon_{-1} = a^{-2}x_0 - a^{-1}\epsilon_{-1} - a^{-2}\epsilon_0$$

$$\vdots$$

$$x_t = a^t x_0 - \sum_{j=t}^{-1} a^j \epsilon_{t-j} \text{ for } t < 0.$$

The solution is unique for the given initial value x_0, but since x_0 is arbitrary, we obtain an infinite number of solutions for $a \neq 0$. We now first suppose that the initial value $x_0 \in \mathbb{R}$ is deterministic. Then we obtain for $s \geq t \geq 0$

$$\mathbf{E}x_t = a^t x_0$$
$$\mathbf{Var}(x_t) = \sigma^2 \sum_{j=0}^{t-1} a^{2j} \tag{1.15}$$
$$\mathbf{Cov}(x_s, x_t) = \mathbf{E}(\sum_{i=0}^{s-1} a^i \epsilon_{s-i})(\sum_{j=0}^{t-1} a^j \epsilon_{t-j}) = \sigma^2 a^{s-t} \sum_{j=0}^{t-1} a^{2j}.$$

Analogous formulas can be derived for the general case $t, s \in \mathbb{Z}$. These solutions are therefore *not* stationary. We can distinguish three essential cases:

(1) For $|a| < 1$, we have

$$\mathbf{E}x_t \longrightarrow 0$$

$$\mathbf{Var}(x_t) = \sigma^2 \sum_{j=0}^{t-1} a^{2j} \longrightarrow \frac{1}{1-a^2}\sigma^2$$

for $t \to \infty$. This case is called *stable case* because both expectation and variance are bounded for all $t \geq 0$.
(2) For $|a| > 1$, however, we have

$$|\mathbf{E}x_t| = |a|^t |x_0| \longrightarrow \infty$$

$$\mathbf{Var}(x_t) = \sigma^2 \sum_{j=0}^{t-1} a^{2j} = \frac{1-a^{2t}}{1-a^2}\sigma^2 \longrightarrow \infty$$

for $t \longrightarrow \infty$. This is the *exponential, unstable* case.
(3) For $|a| = 1$, the following holds:

$$|\mathbf{E}x_t| = |x_0|$$
$$\mathbf{Var}(x_t) = \sigma^2 t.$$

Here, too, the variance increases unboundedly with t, but only at a "linear rate". For $a = 1$, in particular, one obtains a solution that is called *random walk*

$$x_t = \sum_{j=0}^{t-1} \epsilon_{t-j} + x_0, \ t \geq 0. \tag{1.16}$$

In order to obtain a stationary solution, one therefore has to suitably choose the random initial value x_0. We can proceed as follows for the stable case $|a| < 1$: We "start" the system at time $t = -T$ with an arbitrary but bounded initial value x_{-T} (i.e. $\|x_{-T}\| < c < \infty$). For $t \geq -T$, we obtain $x_t = a^{t+T} x_{-T} + \sum_{j=0}^{t+T-1} a^j \epsilon_{t-j}$ by way of recursive insertion, in analogy to above. Now we consider the limit for $T \to \infty$

$$x_t^o := \underset{T \to \infty}{\text{l.i.m}} \left(a^{t+T} x_{-T} + \sum_{j=0}^{t+T-1} a^j \epsilon_{t-j} \right) = \sum_{j=0}^{\infty} a^j \epsilon_{t-j}. \tag{1.17}$$

The sum on the right exists because the coefficients are square summable ($\sum_{j=0}^{\infty} a^{2j} = (1 - a^2)^{-1} < \infty$). Thus, the process (x_t^o) is a causal MA(∞) process and hence stationary. Now (x_t^o) is a solution, since

$$x_t^o = \sum_{j \geq 0} a^j \epsilon_{t-j} = \epsilon_t + \sum_{j \geq 1} a^j \epsilon_{t-j} = \epsilon_t + a \sum_{j \geq 0} a^j \epsilon_{t-1-j} = \epsilon_t + ax_{t-1}^o.$$

This solution is called *steady-state* solution, because it is obtained by "starting the system in the infinite past". The covariance function $\gamma(k)$ of (x_t^o) is, according to (1.12), of the form

$$\gamma(k) = \mathbf{E} x_{t+k}^o x_t^o = \sigma^2 \sum_{j \geq \min(0, -k)} a^{j+k} a^j = \sigma^2 \frac{a^{|k|}}{1 - a^2}. \tag{1.18}$$

Hence, AR(1) processes have a memory that decays at a geometric rate. Correlations are positive for $a > 0$ (Fig. 1.2).

In conclusion of this discussion of the AR(1) case, we would like to note that the solutions of the AR(1) system (1.14) can be written as a sum of a particular solution (i.e. in the stable case, e.g. (x_t^o)) and as a solution of the homogeneous system

$$x_t - ax_{t-1} = 0.$$

This observation can be used to show that in the stable case

(1) every (square integrable) solution for $t \to \infty$ converges to x_t^o (that is, more precisely $\text{l.i.m}_{t \to \infty}(x_t - x_t^o) = 0$) and
(2) the steady-state solution (x_t^o) is the only stationary solution.

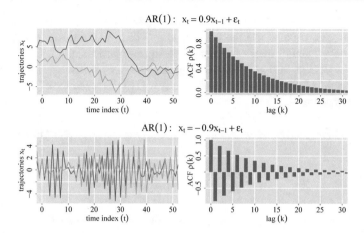

Fig. 1.2 The figure shows two trajectories and the autocorrelation function of the two AR(1) processes $x_t = ax_{t-1} + \epsilon_t$ with $a = 0.9$ and $a = -0.9$

ARMA Processes

An *ARMA system* (<u>A</u>uto<u>r</u>egressive <u>M</u>oving <u>A</u>verage System) is a difference equation of the form

$$x_t = a_1 x_{t-1} + \cdots + a_p x_{t-p} + \epsilon_t + b_1 \epsilon_{t-1} + \cdots + b_q \epsilon_{t-q} \qquad (1.19)$$

with coefficients a_j, $b_j \in \mathbb{R}^{n \times n}$ and white noise $(\epsilon_t) \sim \mathrm{WN}(\Sigma)$. A stationary solution of such a system is called *ARMA process*. We will discuss ARMA systems and processes in Chap. 6.

Harmonic Processes

Harmonic processes are defined by the superposition of (finitely many) harmonic oscillations with stochastic amplitudes and phases. They are not important for applications, but they are very important for the interpretation of general stationary processes, since any stationary process can be approximated with arbitrary accuracy (pointwise in t) by harmonic processes. This will be discussed in more detail in Chap. 3. Also, harmonic processes are examples of so-called *singular* processes, that is, processes that can be predicted exactly. We will have a closer look at them in Chap. 2.

As harmonic oscillations can be represented more elegantly with complex numbers, we consider complex-valued stochastic processes here. Moreover, we consistently use for complex matrices $a = (a_{ij}) \in \mathbb{C}^{m \times n}$ the notation $\bar{a} = (\bar{a}_{ij})$ for the complex conjugate matrix and $a^* = (\bar{a})' \in \mathbb{C}^{n \times m}$ for the Hermitian transpose matrix. A complex square matrix $a \in \mathbb{C}^{n \times n}$ is called *positive semidefinite*, if $x a x^* \geq 0$ holds for all $x \in \mathbb{C}^{1 \times n}$. It is called *positive definite*, if $x a x^* > 0$ for all $x \neq 0 \in \mathbb{C}^{1 \times n}$. As in the real case, we also use the notation $a \geq 0$ ($a > 0$) for positive semidefinite (resp. positive definite) matrices and for two square matrices $a, b \in \mathbb{C}^{n \times n}$, ($a \geq b$) (resp. $a > b$) means that $(a - b) \geq 0$ (resp. $(a - b) > 0$) holds.

For the sake of simplicity, we restrict ourselves to the scalar case ($n = 1$) in this introductory section.

Definition 1.24 (*harmonic processes*) A (scalar) *harmonic process* is a process of the form

$$x_t = \sum_{k=1}^{K} z_k \exp(i \lambda_k t), \quad \text{for } t \in \mathbb{Z} \tag{1.20}$$

where $-\pi < \lambda_1 < \lambda_2 < \cdots < \lambda_K \leq \pi$ and the z_k's are complex-valued random variables.

Since here harmonic oscillations $e^{i\lambda t}$ are observed only for $t \in \mathbb{Z}$, we can restrict ourselves to angular frequencies λ in the interval $(-\pi, \pi]$, because $e^{i\lambda t} = e^{i(\lambda + 2\pi)t}$ holds for each $\lambda \in \mathbb{R}$ and for all $t \in \mathbb{Z}$. The maximum observable frequency $f = \frac{\pi}{2\pi} = \frac{1}{2}$ is called *Nyquist* frequency. However, as the following proposition shows, one has to impose some conditions on frequencies (λ_k) and amplitudes (z_k) for the process (x_t) to be real-valued and stationary. It is convenient to use the following alternative representation for this analysis:

$$x_t = \sum_{m=-M+1}^{M} z_m \exp(i \lambda_m t), \quad 0 = \lambda_0 < \lambda_1 < \cdots < \lambda_M = \pi \text{ and } \lambda_{-m} = -\lambda_m.$$
$$\tag{1.21}$$

This means that we complement (if necessary) the original set of frequencies $\{\lambda_k\}$ with "mirrored" frequencies $-\lambda_k$ and frequencies 0 and π. The corresponding amplitudes of the complemented frequencies are simply set equal to zero. In addition, we change the indexing from $k = 1, \ldots, K$ to $m = 1 - M, \ldots, M$. Altogether, we now have $K = 2M$ frequencies, although some amplitudes may be equal to zero.

Proposition 1.25 (harmonic processes) *A harmonic process* (1.21) *is a real-valued weakly stationary process if and only if the following conditions are satisfied:*

(1) The random variables z_m are square integrable (i.e. $\mathbb{E}|z_m|^2 < \infty$).
(2) $\mathbb{E}z_m = 0$ for $m \neq 0$.
(3) $\mathbb{E}z_m \bar{z}_l = 0$ for all $m \neq l$.
(4) $z_{-m} = \overline{z_m}$ for $0 < m < M$, $z_0 = \overline{z_0}$ and $z_M = \overline{z_M}$.

If these conditions are satisfied, the following holds:

$$\mathbf{E}x_t = \mathbf{E}z_0 \tag{1.22}$$

$$\gamma(k) = \mathbf{Cov}(x_{t+k}, x_t) = \left(\sum_{m=1-M}^{M} \mathbf{E}|z_m|^2 \exp(i\lambda_m k) \right) - (\mathbf{E}z_0)^2. \tag{1.23}$$

Proof We first define the random vector $z = (z_{1-M}, \ldots, z_M)'$ and $\theta_m = \exp(i\lambda_m)$ for $m = 1 - M, \ldots, M$. Thus, we have $|\theta_m| = 1$, $\theta_{-m} = \overline{\theta_m}$ and $\theta_0, \theta_M \in \mathbb{R}$. The random vector $x_0^K = (x_0, x_{-1}, \ldots, x_{-K+1})'$ can be represented as

$$x_0^K = \Theta z$$

where

$$\Theta = \begin{pmatrix} 1 & 1 & \cdots & 1 \\ \theta_{1-M}^{-1} & \theta_{2-M}^{-1} & \cdots & \theta_M^{-1} \\ \vdots & \vdots & & \vdots \\ \theta_{1-M}^{1-K} & \theta_{2-M}^{1-K} & \cdots & \theta_M^{1-K} \end{pmatrix} \in \mathbb{C}^{K \times K}.$$

Matrix Θ is a Vandermonde matrix and is therefore non singular (because of $\theta_k \neq \theta_l$ for $k \neq l$). Hence, we also have $z = \Theta^{-1} x_0^K$. This means that the components of z are square integrable if and only if the x_t's are square integrable.

It is easy to see that the random variables x_t are real-valued if the conditions of point (4) are satisfied. The necessity of point (4) follows from

$$0 = x_0^K - \overline{x_0^K} = \Theta z - \overline{\Theta} \overline{z} = \Theta(z - S\overline{z})$$

where $S \in \mathbb{R}^{K \times K}$ is the permutation matrix defined by $\overline{\Theta} = \Theta S$. Due to the fact that Θ is non singular, it must hold that

$$z - S\overline{z} = (z_{-M+1}, \ldots, z_{-1}, z_0, z_1, \ldots, z_{M-1}, z_M)' -$$
$$(\overline{z_{M-1}}, \ldots, \overline{z_1}, \overline{z_0}, \overline{z_{-1}}, \ldots, \overline{z_{1-M}}, \overline{z_M})' = 0.$$

The expectation $\mathbf{E}x_t = \sum_m \mathbf{E}z_m \exp(i\lambda_m t)$ is constant (independent of t), if $\mathbf{E}z_m = 0$ holds for all $m \neq 0$. Conversely, it follows from

$$\mathbf{E}x_0^K = (1, \ldots, 1)' \mathbf{E}x_0 = \Theta \mathbf{E}z$$

that $\mathbf{E}z = \Theta^{-1}(1, \ldots, 1)' \mathbf{E}x_0 = (0, \ldots, 0, 1, 0, \ldots 0)' \mathbf{E}x_0$, i.e. $\mathbf{E}z_m = 0$ for $m \neq 0$.

We immediately obtain the representation (1.23) of the autocovariance function $\gamma(k)$ of (x_t) if points (1-3) are satisfied. Now suppose that (x_t) is weakly stationary, then

$$\Gamma_K = \mathbf{E}x_0^K (x_0^K)^* = \mathbf{E}x_1^K (x_1^K)^*.$$

The two random vectors x_0^K and x_1^K are represented as $x_0^K = \Theta z$ and $x_1^K = \Theta \mathrm{diag}(\theta_{1-M}, \ldots, \theta_M) z$. From this, it follows that

$$\Gamma_K = \Theta(\mathbf{E}zz^*)\Theta^* = \Theta \mathrm{diag}(\theta_{1-M}, \ldots, \theta_M)(\mathbf{E}zz^*)\mathrm{diag}(\overline{\theta_{1-M}}, \ldots, \overline{\theta_M})\Theta^*$$

and respectively

$$\Theta^{-1}\Gamma_K\Theta^{-*} = \mathbf{E}zz^* = \mathrm{diag}(\theta_{1-M}, \ldots, \theta_M)(\mathbf{E}zz^*)\mathrm{diag}(\overline{\theta_{1-M}}, \ldots, \overline{\theta_M}).$$

It now follows for $m \neq l$, from $\mathbf{E}z_m\overline{z_l} = \theta_m\overline{\theta_l}\mathbf{E}z_m\overline{z_l}$ and $\theta_m \neq \theta_l$, that $\mathbf{E}z_m\overline{z_l} = 0$. $\square$

Exercise 1.26 Let $z_k = a_k + ib_k, k = 1, 2$ be two complex-valued random variables (with $\Re(z_k) = a_k$ and $\Im(z_k) = b_k$). Show $z_1 = \overline{z_2}$ and $\mathbf{E}z_1\overline{z_2} = 0$ is equivalent to $a_2 = a_1, b_2 = -b_1, \mathbf{E}a_1^2 = \mathbf{E}b_1^2$ and $\mathbf{E}a_1b_1 = 0$.

Exercise 1.27 Let (x_t) be a stationary, real-valued process of the form (1.21). Show that (x_t) also has the following representation:

$$x_t = a_0 + a_M(-1)^t + \sum_{m=1}^{M-1} (a_m \cos(\lambda_m t) + b_m \sin(\lambda_m t)) \tag{1.24}$$

where the (real) random variables $a_0, \ldots, a_M, b_1, \ldots b_{M-1}$ satisfy the following conditions: $\mathbf{E}a_m = \mathbf{E}b_m = 0$ for $m > 0$, $\mathbf{E}a_m^2 = \mathbf{E}b_m^2$ for $1 \leq m < M$ and all random variables are uncorrelated (orthogonal) to each other. Note: Set $z_0 = a_0$, $z_M = a_M$ and $z_m = \frac{1}{2}(a_m - ib_m)$ for $1 \leq m < M$.

Exercise 1.28 Given is a process of the form $(x_t = a \cos(\lambda t + \phi) \mid t \in \mathbb{Z})$ with $0 < \lambda < \pi$ and two real-valued, independent random variables a and ϕ. The random variable a is square integrable and $\mathbf{E}a \neq 0$ and $\mathbf{E}a^2 > 0$ holds. Prove that (x_t) is (weakly) stationary if and only if it holds that

$$\mathbf{E}\sin(\phi) = \mathbf{E}\cos(\phi) = \mathbf{E}\sin(2\phi) = \mathbf{E}\cos(2\phi) = 0.$$

Note: Write the process as $x_t = z_1 \exp(-i\lambda t) + z_2 \exp(i\lambda t)$ with suitably chosen complex-valued random variables z_1 and z_2 and then use Theorem 1.25.

Exercise 1.29 Show that expectation and autocovariance function of a real-valued, stationary harmonic process (x_t) of the form (1.24) are given by

$$\mathbf{E}x_t = \mathbf{E}a_0$$

$$\gamma(k) = \mathbf{Cov}(x_{t+k}, x_t) = \mathbf{Var}(a_0) + \mathbf{E}a_M^2(-1)^k + \sum_{m=1}^{M-1} \mathbf{E}a_m^2 \cos(\lambda_m k).$$

Exercise 1.30 Show that a stationary process is a harmonic process if and only if its time domain $\mathbb{H}(x)$ is finite-dimensional. It follows directly from the representation of the process that the time domain of a harmonic process is finite-dimensional. To show the other direction, one can use the eigenvectors and eigenvalues of the forward shift operator U.

We now consider a stationary, real-valued harmonic process (x_t) as defined in (1.21) and we assume that $\mathbf{E}x_t = \mathbf{E}z_0 = 0$. Then we define a process $(z(\lambda) \mid \lambda \in [-\pi, \pi])$ by

$$z(\lambda) = \sum_{\{m \mid \lambda_m \leq \lambda\}} z_m \tag{1.25}$$

and a function $F: [-\pi, \pi] \longrightarrow \mathbb{R}$ by

$$F(\lambda) = \mathbf{E}|z(\lambda)|^2 = \sum_{\{m \mid \lambda_m \leq \lambda\}} \mathbf{E}|z_m|^2. \tag{1.26}$$

The function F is a monotonically nondecreasing, right-continuous step function (with $F(-\pi) = 0$ and $F(\pi) = \mathbf{E}|z(\pi)|^2 = \mathbf{E}x_t^2 < \infty$) and hence defines a discrete measure on the interval $[-\pi, \pi]$. One can easily see that the autocovariance function $\gamma(k)$ of the process (x_t) can be represented as follows:

$$\mathbf{E}x_{t+k}x_t = \gamma(k) = \int_{-\pi}^{\pi} \exp(i\lambda k)dF(\lambda). \tag{1.27}$$

This so-called *spectral distribution function F* is in one-to-one relation with the covariance function γ and can be interpreted as follows: The points of discontinuity of F mark the existing frequencies and the sizes of the jumps $(\mathbf{E}|z_m|^2)$ are a measure for the magnitude of the amplitudes (z_m) and are thus (in a physical interpretation) a measure for the expected power of the oscillation components (Fig. 1.3).

Also, the process itself has a corresponding *spectral representation* (or Fourier representation)

$$x_t = \int_{-\pi}^{\pi} \exp(i\lambda t)dz(\lambda). \tag{1.28}$$

In Chap. 3, we will discuss how to interpret this *stochastic integral* (1.28) in a more detailed way. We shall see, in particular, that every stationary process has such a spectral representation (1.28) and that the autocovariance function also always has a representation as given in Eq. (1.27).

1.5 Examples of Non-stationary Processes

In practical applications of time series analysis, one often considers classes of non-stationary processes that are either based on stationary processes or which can be

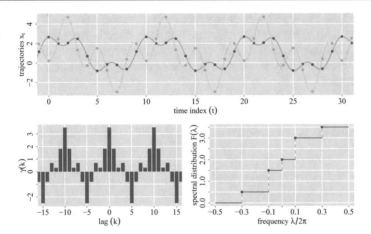

Fig. 1.3 The figure shows two trajectories, the autocovariance function $\gamma(k)$ and the spectral distribution function $F(\lambda)$ of a harmonic process. The trajectories here are plotted for $t \in \mathbb{R}$ in order to visualize the nature of the harmonic oscillations

reduced to stationary processes by suitable transformations. This is another reason why the theory of stationary processes is of importance. Two elementary examples are briefly presented here.

A simple example of a non-stationary process is

$$x_t = \mu_t + u_t$$

where (u_t) is stationary with a mean equal to zero and $\mu_t \not\equiv$ const denotes a deterministic function of time. Thus, the process is the superposition of a deterministic trend (μ_t) and a stationary process (u_t). In the simplest case, $\mu_t = \alpha + \beta t$ is a linear trend.

A *random walk* is a process of the form

$$\left(x_t = \sum_{k=1}^{t} \epsilon_k \,\middle|\, t \in \mathbb{N} \right)$$

where $(\epsilon_t) \sim \mathrm{WN}(\sigma^2)$ is (scalar) white noise. See also the discussion of the AR(1) case. Note that we here consider processes with an index set $\mathbb{N}$ rather than $\mathbb{Z}$. The definition of stationarity obviously can be carried over to this case.

A *random walk with drift* is defined by

$$\left(x_t = \alpha + \beta t + \sum_{k=1}^{t} \epsilon_k \,\middle|\, t \in \mathbb{N} \right).$$

The moments of x_t are

$$\mathbf{E}x_t = \alpha + \beta t$$

$$\mathbf{Cov}(x_t, x_s) = \min(t, s)\sigma^2$$

and (x_t) is hence non-stationary. The process (x_t) is a solution of the difference equation

$$x_t - x_{t-1} = \beta + \epsilon_t, \quad t \in \mathbb{N}$$

if one sets $x_0 = \alpha$.

In more general terms, one often considers processes that are solutions of the difference equation

$$x_t = x_{t-1} + u_t, \quad t \in \mathbb{N}$$

where (u_t) denotes a stationary process. The solutions of this difference equation are

$$x_t = x_0 + \sum_{j=1}^{t} u_j.$$

Typically, (x_t) is non-stationary. However, the first differences $(x_t - x_{t-1} = u_t)$ are stationary by construction. Non-stationary processes with this property are called *integrated of order one (I(1) process)*. Correspondingly, a stationary process is sometimes also called integrated of order zero, or $I(0)$ process, respectively.

Example

However, consider the following example: Suppose $(u_t = v_t - v_{t-1})$, where (v_t) is stationary. The process

$$x_t = x_0 + \sum_{j=1}^{t} u_j = x_0 + v_t - v_0$$

is *stationary* if we choose $x_0 = v_0$ as initial value.

References

M. Bolla, T. Szabados, *Multidimensional Stationary Time Series: Dimension Reduction and Prediction*, 1st edn. (Chapman and Hall/CRC, 2021)

P. Brockwell, R. Davis, *Time Series: Theory and Methods*, Springer Series in Statistics, 2nd edn. (Springer, New York, 1991)

M. Brokate, G. Kersting, *Maß und Integral* (Mathematik Kompakt. Springer Basel AG, Basel, 2011). 9783034606462

J.L. Doob, *Stochastic Processes* (Wiley, 1953)

M. Pourahmadi, *Foundations of Time Series Analysis and Prediction Theory* (Wiley, 2001)

K.D. Schmidt, *Maß und Wahrscheinlichkeit* (Springer-Lehrbuch. Springer, Berlin, Heidelberg, 2009). 978-3-540-89729-3

Prediction

2

The computation of "good" predictions and a quantitative analysis of the prediction quality are among the most important applications of time series analysis. Prediction in general is concerned with approximating a future process variable x_{t+h} by a function

$$\hat{x}_{t,h} = g(x_t, x_{t-1}, \ldots)$$

of the observed values up to the present time t, where $h > 0$ is the so-called *prediction horizon*. Typically this function is chosen from a class of functions such that the prediction error is minimal. For an exact formulation of the problem, one needs to specify the class of functions, i.e. the set of such prediction functions $g(\cdot)$, as well as a measure of the quality of approximation. Here, we discuss a special prediction problem, namely the so-called linear least squares prediction, which means that we restrict ourselves to *linear* (and more precisely, *affine*) prediction functions, i.e. functions of the form

$$\hat{x}_{t,h} = g(x_t, x_{t-1}, \ldots) = c_0 + c_1 x_t + c_2 x_{t-1} + \cdots$$

and to the *mean squared error* (MSE)

$$\mathbf{E}(x_{t+h} - \hat{x}_{t,h})'(x_{t+h} - \hat{x}_{t,h})$$

as quality criterion. We consider an idealized problem for which we assume that we know the exact properties of the underlying process (expectation and covariance function). This idealized prediction problem can be handled simply and elegantly by using the projection theorem. In order to obtain a "feasible" prediction, we first need to estimate the population moments based on data.

© Springer Nature Switzerland AG 2022
M. Deistler and W. Scherrer, *Time Series Models*, Lecture Notes in Statistics 224,
https://doi.org/10.1007/978-3-031-13213-1_2

Without the restriction to linear functions, the conditional expectation

$$\mathbf{E}\left[x_{t+h} \mid x_t, x_{t-1}, \dots\right]$$

is the optimal least squares approximation of x_{t+h} based on past values. However, the calculation of this conditional expectation requires knowledge of the joint distribution of the observed random variables and is therefore often difficult to compute or estimate in practice.

Despite the fact that the restriction to linear predictions and the quadratic performance criterion is a limitation, this prediction is the one that is used most often. In cases where it is possible to assign costs to mispredictions, it may be reasonable to use these costs as the loss function for optimizing the prediction. In particular, this is indicated if these costs are different for "under-" and "overprediction", respectively, by the same amount.

The first section of this chapter uses the last k observed values for prediction. Hence, one also speaks of prediction based on a finite past. The transition $k \to \infty$, which is accordingly called prediction based on the infinite past, subsequently leads to the so-called Wold decomposition of stationary processes. See Wold (1954), Kolmogorov (1941). This Wold decomposition is important for prediction and, moreover, crucial for the understanding of the structure of stationary processes.

The prediction problem for scalar stationary processes has been completely solved by Kolmogorov (see, e.g. Rozanov 1967, Chaps. 1–3). For the multidimensional case, we refer to Rozanov (1967)[1] and Hannan (1970).[2]

2.1   Prediction from a Finite Past

We now want to construct the optimal linear h-step ahead prediction from a finite past. The corresponding optimization problem

$$\mathbf{E}(x_{t+h} - \check{x})'(x_{t+h} - \check{x}) \longrightarrow \min$$
$$\check{x} = c_0 + c_1 x_t + \cdots + c_k x_{t+1-k}$$

can be decomposed into n independent subproblems

$$\mathbf{E}(x_{i,t+h} - \check{x}_i)^2 \longrightarrow \min$$
$$\check{x}_i = c_{i0} + c_{i1} x_t + \cdots + c_{ik} x_{t+1-k}$$

where $c_{i0} \in \mathbb{R}$ denotes the ith element of c_0 and $c_{ij} \in \mathbb{R}^{1 \times n}$ denotes the ith row of $c_j \in \mathbb{R}^{n \times n}$ ($i = 1, \dots, n$). The equivalent formulation in the Hilbert space of square integrable random variables $\mathbb{L}_2(\Omega, \mathcal{A}, \mathbf{P})$ is as follows:

---

[1] Yuri A. Rozanov. Russian mathematician, who has written a standard book on stationary processes, former student of Kolmogorov.
[2] Edward J. Hannan (1921–1994). Australian statistician and pioneer of modern time series analysis.

$$\|x_{i,t+h} - \check{x}_i\| \longrightarrow \min$$
$$\check{x}_i \in \mathrm{sp}\{1, x_t, \ldots, x_{t+1-k}\} =: \mathbb{M} \subset \mathbb{L}_2.$$

Here again, the 1 is to be interpreted as a random variable ($\omega \mapsto 1$). The solution follows directly from the projection Theorem 1.11

$$\hat{x}_{i,t,h,k} = P_{\mathbb{M}} x_{i,t+h}. \tag{2.1}$$

Thus, the optimal prediction for $x_{i,t+h}$ is the projection of $x_{i,t+h}$ onto the subspace $\mathbb{M}$. If we use the convention for random vectors introduced in Sect. 1.3, we can also write

$$\hat{x}_{t,h,k} = P_{\mathbb{M}} x_{t+h}. \tag{2.2}$$

We will now show that we can essentially restrict ourselves to centered processes (i.e. $\mathbf{E}x_t = 0$) and linear predictions (i.e. $c_0 = 0$). For this purpose, we define the mean-adjusted process ($\tilde{x}_t = x_t - \mu$) with $\mu = \mathbf{E}x_t$. The subspace $\mathbb{M} = \mathrm{sp}\{1, x_t, \ldots, x_{t+1-k}\}$ is the direct sum of the two *orthogonal* subspaces $\mathrm{sp}\{1\}$ and $\tilde{\mathbb{M}} := \mathrm{sp}\{\tilde{x}_1, \ldots, \tilde{x}_{t+1-k}\}$, since $\langle 1, \tilde{x}_{is} \rangle = \mathbf{E}\tilde{x}_{is} = 0$. Hence, the projection onto $\mathbb{M}$ is equal to the sum of the projections onto $\mathrm{sp}\{1\}$ and the projection onto $\tilde{\mathbb{M}}$, that is, $P_{\mathbb{M}} = P_{\mathrm{sp}\{1\}} + P_{\tilde{\mathbb{M}}}$. Now, using the linearity of projection operators, it further follows that

$$P_{\mathbb{M}} x_{t+h} = (P_{\mathrm{sp}\{1\}} + P_{\tilde{\mathbb{M}}})(\tilde{x}_{t+h} + \mu)$$
$$= P_{\mathrm{sp}\{1\}} \tilde{x}_{t+h} + P_{\mathrm{sp}\{1\}} \mu + P_{\tilde{\mathbb{M}}} \tilde{x}_{t+h} + P_{\tilde{\mathbb{M}}} \mu = \mu + P_{\tilde{\mathbb{M}}} \tilde{x}_{t+h}$$

since $\tilde{x}_{i,t+h} \perp \mathrm{sp}\{1\}$, $\mu_i 1 \perp \tilde{\mathbb{M}}$ and $\mu_i 1 \in \mathrm{sp}\{1\}$. This shows the following:

(1) The optimal prediction of the centered process is linear (i.e. $c_0 = 0$):

$$\hat{\tilde{x}}_{t,h,k} = (P_{\mathrm{sp}\{1\}} + P_{\tilde{\mathbb{M}}})\tilde{x}_{t+h} = P_{\tilde{\mathbb{M}}} \tilde{x}_{t+h} = c_1 \tilde{x}_t + \cdots + c_k \tilde{x}_{t+1-k}.$$

(2) The prediction for x_{t+h} is obtained simply by adding the expectation $\mu = \mathbf{E}x_{t+h}$ to the prediction of the centered process:

$$\hat{x}_{t,h,k} = \mu + \hat{\tilde{x}}_{t,h,k} = (I_n - c_1 - \cdots - c_k)\mu + c_1 x_t + \cdots + c_k x_{t+1-k}.$$

Similar considerations apply to predictions from the infinite past. We will therefore assume in the following, without loss of generality, that the process under consideration is already centered. Therefore, we shall consider only linear predictions ($c_0 = 0$). Accordingly, $\mathbb{M}$ from now on denotes the subspace $\mathbb{M} = \mathrm{sp}\{x_t, \ldots, x_{t+1-k}\}$.

The projection theorem provides a linear system of equations to determine the prediction coefficients. A linear combination $c_{i1} x_t + \cdots + c_{ik} x_{t+1-k} \in \mathbb{M}$ is equal to the projection of $x_{i,t+h}$ onto space $\mathbb{M}$ if and only if the error $x_{i,t+h} - (c_{i1}x_t + \cdots + c_{ik}x_{t+1-k})$ is orthogonal to this space, that is, if and only if

$$\langle x_{i,t+h} - c_{i1}x_t - \cdots - c_{ik}x_{t+1-k}, x_{j,t+1-l} \rangle = 0 \\ \text{for } 1 \leq j \leq n \text{ and } 1 \leq l \leq k. \tag{2.3}$$

Summarizing Eq. (2.3) with $\hat{x}_{t,h,k} = (c_1, \ldots, c_k)x_t^k$ to

$$\mathbf{E}\left[(x_{t+h} - (c_1, c_2, \ldots, c_k)x_t^k)(x_t^k)'\right] = 0$$

leads to the following "prediction equations":

$$(\gamma(h), \gamma(h+1), \ldots, \gamma(h+k-1)) = (c_1, \ldots, c_k)\Gamma_k \tag{2.4}$$

for determining the coefficients $(c_1, \ldots, c_k)$.

The (optimal) prediction error $u_{t,h,k} = x_{t+h} - \hat{x}_{t,h,k} = x_{t+h} - (c_1, \ldots, c_k)x_t^k$ has expected value equal to zero. Here we use the stacked vector x_t^k introduced in (1.3). The variance $\Sigma_{h,k}$ of the prediction error can be determined as follows: Since $\hat{x}_{j,t,h,k} \in \mathbb{M}$, it follows that $\langle u_{i,t,h,k}, \hat{x}_{j,t,h,k} \rangle = \mathbf{Cov}(u_{i,t,h,k}, \hat{x}_{j,t,h,k}) = 0$ and hence $\mathbf{Var}(x_{t+h}) = \mathbf{Var}(\hat{x}_{t,h,k} + u_{t,h,k}) = \mathbf{Var}(\hat{x}_{t,h,k}) + \mathbf{Var}(u_{t,h,k})$. Thus, we have

$$\Sigma_{h,k} := \mathbf{E}u_{t,h,k}u'_{t,h,k} = \mathbf{Var}(x_{t+h}) - \mathbf{Var}(\hat{x}_{t,h,k}) = \\ = \gamma(0) - (c_1, \ldots, c_k)\Gamma_k(c_1, \ldots, c_k)'. \tag{2.5}$$

The mean squared error (MSE) of the optimal prediction equals

$$\mathbf{E}\left[u'_{t,h,k}u_{t,h,k}\right] = \text{tr}(\Sigma_{h,k}).$$

We can draw the following conclusions from the projection theorem: The (optimal) prediction $\hat{x}_{t,h,k}$ and hence the corresponding prediction error $u_{t,h,k}$ (and its variance $\Sigma_{h,k}$) are (almost surely) unique. The prediction equations, see (2.4), are always solvable, even if the matrix Γ_k is singular.

If $\Gamma_k > 0$ is positive definite, then (2.4) has a unique solution

$$(c_1, \ldots, c_k) = (\gamma(h), \ldots, \gamma(h+k-1))\Gamma_k^{-1} \tag{2.6}$$

and the prediction error variance can be calculated as

$$\Sigma_{h,k} = \gamma(0) - (\gamma(h), \ldots, \gamma(h+k-1))\Gamma_k^{-1}(\gamma(h), \ldots, \gamma(h+k-1))'. \tag{2.7}$$

Infinitely many solutions exist if Γ_k is singular. In this case, the random variables $\{x_{1t}, \ldots, x_{nt}, x_{1,t-1}, \ldots, x_{n,t+1-k}\}$ are linearly dependent and thus do not constitute a basis for $\mathbb{M}$. See also Exercise 1.13.

Due to the linearity of the projector, we can also immediately derive the optimal prediction for arbitrary linear combinations cx_{t+h}, $c \in \mathbb{R}^{1 \times n}$ as

$$P_M(cx_{t+h}) = c(P_M x_{t+h}) = c\hat{x}_{t,h}.$$

Thus,

$$\Sigma_{h,k} \leq \mathbf{E}(x_{t+h} - \tilde{x})(x_{t+h} - \tilde{x})' \tag{2.8}$$

follows for all $\tilde{x}$ of the form $\tilde{x} = \tilde{c}_0 + \tilde{c}_1 x_t + \cdots + \tilde{c}_k x_{t+1-k}, \tilde{c}_0 \in \mathbb{R}^{n \times 1}, \tilde{c}_i \in \mathbb{R}^{n \times n}$ $i = 1, \ldots, k$. The prediction is hence also optimal with respect to the partial ordering "$\geq$". Another conclusion is

$$\Sigma_{h,k} \leq \Sigma_{h,k-1}, \tag{2.9}$$

that is, the prediction becomes better in general (at least it cannot become worse), if more information is available for prediction. (This inequality follows from (2.8), if one sets $\tilde{x} = \hat{x}_{t,h,k-1}$.)

If $\Sigma_{h,k}$ is singular, we have linear combinations cx_{t+h} that are predicted perfectly (i.e. without error). In this case, the (block) Toeplitz matrix Γ_{k+h} is singular as well.

In the following, we will refer to these prediction(s) as linear least squares prediction(s). Analogous considerations also apply to non-stationary (but square integrable) processes. The only difference is that the variance matrix $\mathbf{Var}(x_t^k)$ no longer needs to have a (block) Toeplitz structure and that the prediction coefficients and the variance of the prediction errors also depend on t in general.

Exercise 2.1 Let (x_t) be a centered stationary process. We now consider the prediction for x_{t+1} from values $x_1, \ldots, x_t$ for $t \in \mathbb{N}_0$. Here, the corresponding predictions and prediction errors are denoted by $\hat{x}_{t+1|t}$ and $u_{t+1|t}$. For $t = 0$, we set $x_{1|0} = 0$ and $u_{1|0} = x_1$. Now show the following:

(1) $\dim(\mathrm{sp}\{x_t, \ldots, x_1\}) = \mathrm{rk}\,\Gamma_t$.
(2)

$$\mathrm{sp}\{x_t, \ldots, x_1\} = \mathrm{sp}\{u_{t|t-1}\} \oplus \mathrm{sp}\{x_{t-1}, \ldots, x_1\}$$
$$= \mathrm{sp}\{u_{t|t-1}\} \oplus \mathrm{sp}\{u_{t-1|t-2}\} \oplus \cdots \oplus \mathrm{sp}\{u_{1|0}\},$$

where the subspaces are mutually orthogonal to each other, e.g. $\mathrm{sp}\{u_{t|t-1}\} \perp \mathrm{sp}\{x_{t-1}, \ldots, x_1\}$.
(3)

$$\mathrm{rk}(\Gamma_{t+1}) = \mathrm{rk}(\Gamma_t) + \mathrm{rk}(\Sigma_{1,t}) \tag{2.10}$$

$$(\Gamma_{t+1} > 0) \iff ((\Gamma_t > 0)\ \text{and}(\Sigma_{1,t} > 0)) \tag{2.11}$$

$$(\det(\Gamma_t) = 0) \implies (\det(\Sigma_{1,t-1}) = 0) \implies (\det(\Sigma_{1,t}) = 0). \tag{2.12}$$

Exercise 2.2 (*Continuation of the exercise above*) We now additionally assume that the process is scalar ($n = 1$) and that the Toeplitz matrix $\Gamma_k > 0$ is non-singular. Now consider the *Cholesky decomposition* of the Toeplitz matrix Γ_k

$$\Gamma_k = DSD'$$

where $D = (d_{ij})_{i,j=1,\ldots,k} \in \mathbb{R}^{k \times k}$ is an *upper* triangular matrix ($d_{ii} = 1$ and $d_{ij} = 0$ for $i > j$) and $S \in \mathbb{R}^{k \times k}$ is a diagonal matrix. The inverse of D is denoted by $C = D^{-1} = (c_{ij})_{i,j=1,\ldots,k}$. $c_{ii} = 1$ and $c_{ij} = 0$ holds for $i > j$. Now show

$$(u_{k|k-1}, u_{k-1|k-2}, \ldots, u_{1|0})' = C(x_k, x_{k-1}, \ldots, x_1)'$$

and $S = \text{diag}(\sigma_{1,k-1}^2, \sigma_{1,k-2}^2, \ldots, \sigma_{1,0}^2)$, where $\sigma_{1,t-1}^2 = \mathbf{E}u_{t|t-1}^2$. Thus, it follows for $1 \le l < k$ that

$$\hat{x}_{t,1,l} = -(c_{k-l,k-l+1}x_t + c_{k-l,k-l+2}x_{t-1} + \cdots + c_{k-l,k}x_{t+1-l}).$$

This exercise could be easily generalized to the multivariate case by using a block Cholesky decomposition.

Exercise 2.3 Let $(x_t = \cos(\lambda t) \mid t \in \mathbb{N})$ be a process where λ is a random variable uniformly distributed on $[-\pi, \pi]$. In Exercise 1.7, one was supposed to show that $\mathbf{E}x_t = 0$ and $\gamma(k) = \mathbf{E}x_{t+k}x_t = 0$ holds for $t + k \ge t \ge 0$. Thus, the optimal *linear* prediction is equal to zero, i.e. $\hat{x}_{t,h,k} = 0$ for $1 \le k < t$. However, this process allows for perfect *non-linear* prediction! Show

$$x_{t+1} = 2x_t x_1 - x_{t-1}$$
$$x_t = 2x_{t-1}x_1 - x_{t-2}$$
$$\text{and hence } x_{t+1} = \frac{x_t^2 - x_{t-1}^2 + x_t x_{t-2}}{x_{t-1}}.$$

Exercise 2.4 We consider the scalar AR(1) process $x_t = ax_{t-1} + \epsilon_t$, with $|a| < 1$ and $(\epsilon_t) \sim \text{WN}(\sigma^2)$. See also (1.17) and (1.18). Using Eqs. (2.4) and (2.5), show that for $k \ge 1$

$$\hat{x}_{t,h,k} = a^h x_t \quad \text{and} \quad \sigma_{h,k}^2 = \sigma^2 \frac{(1 - a^{2h})}{(1 - a^2)}.$$

Exercise 2.5 Consider the MA(1) process $x_t = \epsilon_t - \epsilon_{t-1}$, where $(\epsilon_t) \sim \text{WN}(\sigma^2)$ is white noise with variance $\mathbf{E}\epsilon_t^2 = \sigma^2$. Prove the following formulas for the one-step ahead prediction $\hat{x}_{t,1,k}$ from k past values and the corresponding prediction error $u_{t,1,k} = x_{t+1} - \hat{x}_{t,1,k}$:

$$\hat{x}_{t,1,k} = \frac{-1}{k+1}(kx_t + (k-1)x_{t-1} + \cdots + 2x_{t+2-k} + 1x_{t+1-k})$$

$$u_{t,1,k} = \frac{1}{k+1}((k+1)\epsilon_{t+1} - \epsilon_t - \epsilon_{t-1} - \cdots - \epsilon_{t+1-k} - \epsilon_{t-k})$$

$$\sigma_{1,k}^2 = \mathbf{E}(u_{t,1,k}^2) = \sigma^2 + \frac{\sigma^2}{k+1} = \sigma^2 \frac{k+2}{k+1}.$$

Also, show that the one-step ahead prediction error for $k \to \infty$ converges to ϵ_{t+1}, i.e.

$$\underset{k\to\infty}{\mathrm{l.i.m}}\, u_{t,1,k} = \epsilon_{t+1}.$$

As already mentioned above, the same strategy can be applied to determine the least squares prediction for non-stationary (but square integrable) processes.

Exercise 2.6 Let (x_t) be a process of the form $x_t = \mu_t + y_t$, where μ_t is a deterministic function of time and (y_t) is a stationary centered process. Show that

$$\hat{x}_{t,h,k} = \mu_{t+h} + \hat{y}_{t,h,k} = \mu_{t+h} + c_1 y_t + \cdots + c_k y_{t+1-k} =$$
$$= (\mu_{t+h} - c_1\mu_t - \cdots - c_k\mu_{t+1-k}) + c_1 x_t + \cdots + c_k x_{t+1-k}$$

is the best affine prediction for x_{t+h} from k past values. Here, $\hat{y}_{t,h,k} = c_1 y_t + \cdots + c_k y_{t+1-k}$ denotes the h-step ahead prediction for y_{t+h}. $x_{t+h} - \hat{x}_{t,h,k} = y_{t+h} - \hat{y}_{t,h,k}$ holds for the prediction error.

Exercise 2.7 Let (y_t) be a centered stationary process and $(x_t \mid t \in \mathbb{N}_0)$ be the integrated process defined by $x_t = x_0 + \sum_{j=1}^t y_j$. We assume that the initial value x_0 is square integrable and uncorrelated to $y_s, s \geq 1$. Show that

$$\hat{x}_{t+h} = x_t + \hat{y}_{t,1,t} + \hat{y}_{t,2,t} + \cdots + \hat{y}_{t,h,t}$$

is the best prediction for x_{t+h} from values $x_0, \ldots, x_t$.

If $(y_t) \sim WN(\sigma^2)$ is white noise (i.e. (x_t) is a random walk), then the "naive prediction" $\hat{x}_{t+h} = x_t$ is the optimal one.

2.2 Prediction from Infinite Past

In this section, we consider the limit of the predictions $\hat{x}_{t,h,k}$ for $k \to \infty$, that is, we use the information from the entire, infinite past for prediction purposes. Prediction from the infinite past reveals certain structural properties of the underlying process, as we shall see in the next section. Again, we consider only the case of centered processes, i.e. $\mathbf{E}x_t = 0$. However, all results can be readily applied to the general case. The following proposition holds.

Proposition 2.8 *The sequence $(\hat{x}_{t,h,k} \mid k \in \mathbb{N})$ converges (in mean square) and the limit is the projection of x_{t+h} onto the space $\mathbb{H}_t(x) = \overline{\mathrm{sp}}\{x_{is} \mid 1 \le i \le n, s \le t\} = \overline{\mathrm{sp}}\{x_s \mid s \le t\}$, that is,*

$$\mathop{\mathrm{l.i.m}}_{k \to \infty} \hat{x}_{t,h,k} = P_{\mathbb{H}_t(x)} x_{t+h} =: \hat{x}_{t,h}.$$

The variance of the corresponding error $u_{t,h} = x_{t+h} - \hat{x}_{t,h}$ is

$$\Sigma_h := \mathbf{Var}(u_{t,h}) = \mathbf{E}u_{t,h}u'_{t,h} = \lim_{k \to \infty} \Sigma_{h,k}.$$

Proof Suppose $\mathbb{H}_{t,k}(x) = \mathrm{sp}\{x_s \mid t+1-k \le s \le t\}$. Then,

$$P_{\mathbb{H}_{t,k}(x)} = P_{\mathbb{H}_{t,k}(x)} P_{\mathbb{H}_t(x)}$$

follows from $\mathbb{H}_{t,k}(x) \subset \mathbb{H}_t(x)$, and hence $\hat{x}_{t,h,k} = P_{\mathbb{H}_{t,k}(x)} \hat{x}_{t,h}$ where $\hat{x}_{t,h} = P_{\mathbb{H}_t(x)} x_{t+h}$. Note that $\hat{x}_{t,h}$ is a limit of finite sums and thus there exists a sequence of random vectors $x^{(k)} \in (\mathbb{H}_{t,k}(x))^n$ such that $\hat{x}_{t,h} = \mathrm{l.i.m}_k x^{(k)}$. Based on the properties of the projection, we finally get

$$\mathbf{E}(\hat{x}_{t,h} - \hat{x}_{t,h,k})(\hat{x}_{t,h} - \hat{x}_{t,h,k})' \le \mathbf{E}(\hat{x}_{t,h} - x^{(k)})(\hat{x}_{t,h} - x^{(k)})'$$

and thus we have the convergence of $\hat{x}_{t,h,k}$ to $\hat{x}_{t,h}$ for $k \to \infty$. $\qquad\square$

Using one-step ahead prediction based on the infinite past, we obtain a decomposition of the process of the form

$$x_{t+1} = \hat{x}_{t,1} + u_{t,1}$$

where $\hat{x}_{t,1}$ is the part of x_{t+1} that is determined by the past, and $u_{t,1}$ is the "unpredictable" part. This is why the one-step ahead prediction errors from the infinite past are called the *innovations* of the process.

Proposition 2.9 *The innovations $(u_t = u_{t-1,1} \mid t \in \mathbb{Z})$ of a stationary process are white noise.*

Proof Clearly, innovations are (weakly) stationary with an expectation equal to zero ($\mathbf{E}u_{t,1} = 0$). Note that $u_{t-1,1} \in \mathbb{H}_t(x)$ and $u_{t-1,1} \perp \mathbb{H}_{t-1}(x)$ holds. Hence $u_{t-1,1} \perp u_{s-1,1} \in \mathbb{H}_s(x) \subset \mathbb{H}_{t-1}(x)$ holds for all $s \le t-1$ and $u_{t,1} \perp u_{s,1}$ thus also holds for all $s \ne t$. $\qquad\square$

The inequality

$$\Sigma_{h+1} \ge \Sigma_h \qquad (2.13)$$

follows directly from $\mathbb{H}_{t-1}(x) \subset \mathbb{H}_t(x)$ and $\hat{x}_{t-1,h+1} = P_{\mathbb{H}_{t-1}(x)} x_{t+h}$ and $\hat{x}_{t,h} = P_{\mathbb{H}_t(x)} x_{t+h}$.

The prediction method discussed here is based on the covariance function of the process. Given that one has a parametric model for the process (such as an AR model or a state-space model), prediction can be simplified to a considerable extent. This will be discussed in further chapters.

2.3 Regular and Singular Processes and Wold Decomposition

By the Wold decomposition every stationary process can be decomposed into a "deterministic" and a "regular" part. Here, "deterministic" means that the future is completely determined by the past, while "regular" means that the infinitely distant past plays no role for the future. If the Wold decomposition of a process is known, one obtains a simple and explicit representation of the prediction of the regular part (for arbitrary $h > 0$).

Definition 2.10 A stationary process (x_t) is called

- *regular (purely non-deterministic)*, if $\text{l.i.m}_{h\to\infty} \hat{x}_{t,h} = 0$ (and thus $\lim_{h\to\infty} \Sigma_h = \text{E}x_t x_t'$) holds. The expectation of a regular process must be zero ($\text{E}x_t = 0$).
- *singular (deterministic)*, if $\Sigma_h = 0$ holds for one $h > 0$ (and hence also for all $h > 0$). A singular process may also be called *deterministic,* because the future is determined from the past.

Exercise 2.11 Show that $\Sigma_{\bar{h}} = 0$ for one $\bar{h} > 0$ also implies $\Sigma_h = 0$ for all $h > 0$.

Exercise 2.12 Show that harmonic processes (see Eq. (1.20)) are singular. Note: Prove $\hat{x}_{t,1,k} = x_{t+1}$ for $k \geq K$.

Examples of regular processes

(1) MA(q) processes are regular, since $\hat{x}_{t,h} = \text{l.i.m}_{k\to\infty} \hat{x}_{t,h,k} = 0$ for $h > q$.
(2) Processes that have a causal MA(∞) representation are regular:
   Let $x_t = \sum_{k\geq 0} b_k \epsilon_{t-k}$ be a causal MA(∞) representation for process (x_t). Since $x_s \in \mathbb{H}_t(\epsilon) \; \forall s \leq t$, it also follows that $\mathbb{H}_t(x) \subset \mathbb{H}_t(\epsilon)$. We now consider the projection $\tilde{x}_{t,h} = P_{\mathbb{H}_t(\epsilon)} x_{t+h}$ of x_{t+h} onto $\mathbb{H}_t(\epsilon)$. This projection can be easily computed because (ϵ_t) is white noise:

$$\tilde{x}_{t,h} = P_{\mathbb{H}_t(\epsilon)} x_{t+h} = \sum_{k\geq 0} b_k P_{\mathbb{H}_t(\epsilon)} \epsilon_{t+h-k} = \sum_{k\geq h} b_k \epsilon_{t+h-k}.$$

For $h \to \infty$, $\tilde{x}_{t,h}$ converges to zero, that is, $\text{l.i.m}_{h\to\infty} \tilde{x}_{t,h} = 0$. As $\mathbb{H}_t(x)$ is a closed subspace of $\mathbb{H}_t(\epsilon)$, it also follows that $\mathbf{E}(x_{t+h} - \hat{x}_{t,h})(x_{t+h} - \hat{x}_{t,h})' \geq \mathbf{E}(x_{t+h} - \tilde{x}_{t,h})(x_{t+h} - \tilde{x}_{t,h})'$ and $\mathbf{E}(\hat{x}_{t,h})(\hat{x}_{t,h})' \leq \mathbf{E}(\tilde{x}_{t,h})(\tilde{x}_{t,h})'$, respectively. Together this yields $\text{l.i.m}_{h\to\infty} \hat{x}_{t,h} = 0$ as claimed.

The following proposition (Wold decomposition) will show that, conversely, every regular process has a causal MA(∞) representation.

Proposition 2.13 (Wold decomposition)

(1) Every (n-dimensional) stationary process (x_t) has a unique decomposition $x_t = y_t + z_t$ with the following properties:

    a. (y_t) is regular and (z_t) is singular.
    b. The processes (y_t) and (z_t) are orthogonal to each other ($\mathbf{E}y_t z_s' = 0$ for all $t, s \in \mathbb{Z}$).
    c. $y_{it} \in \text{sp}\{1\} + \mathbb{H}_t(x)$ and $z_{it} \in \text{sp}\{1\} + \mathbb{H}_t(x)$ for $i = 1, \ldots, n$.

(2) The regular process (y_t) has a causal MA(∞) representation

$$y_t = \sum_{j \geq 0} b_j \epsilon_{t-j}, \quad \text{with } b_0 = I \text{ and } \sum_{j \geq 0} \|b_j\|^2 < \infty \qquad (2.14)$$

where (ϵ_t) is white noise, $(\mathbb{H}_t(\epsilon) = \mathbb{H}_t(y))$ holds and the ϵ_t's are innovations of both (x_t) as well as (y_t).

Proof For the purpose of making the proof slightly easier, let us assume that $\mathbf{E}x_t = 0$. For the general case, we simply consider the Wold decomposition $\tilde{x}_t = \tilde{y}_t + \tilde{z}_t$ of the centered process $\tilde{x}_t = x_t - \mathbf{E}x_t$ and set $y_t = \tilde{y}_t$ and $z_t = \tilde{z}_t + \mathbf{E}x_t$.

First, we define the innovations

$$\epsilon_t = x_t - P_{\mathbb{H}_{t-1}(x)} x_t \qquad (2.15)$$

of the process (x_t) and note that

$$\epsilon_t \in (\mathbb{H}_t(x))^n \qquad (2.16)$$
$$\mathbb{H}_t(\epsilon) \subset \mathbb{H}_t(x)$$
$$\epsilon_t \perp \mathbb{H}_{t-1}(x).$$

We define the processes (y_t) and (z_t) by

$$y_t = P_{\mathbb{H}_t(\epsilon)} x_t \qquad (2.17)$$
$$z_t = x_t - y_t. \qquad (2.18)$$

The process (ϵ_t) is white noise (see Proposition 2.9); thus, $y_t \in (\mathbb{H}_t(\epsilon))^n$ has the causal MA(∞) representation

$$y_t = \sum_{j \geq 0} b_j \epsilon_{t-j}$$

with square summable coefficients $(b_j)_{j \geq 0}$. Due to

$$b_0 \epsilon_t = P_{sp\{\epsilon_t\}} x_t = P_{sp\{\epsilon_t\}} \epsilon_t + P_{sp\{\epsilon_t\}} P_{\mathbb{H}_{t-1}(x)} x_t = \epsilon_t,$$

we can set $b_0 = I$ without loss of generality. We also see that

$$y_t \in (\mathbb{H}_t(\epsilon))^n$$
$$\mathbb{H}_t(y) \subset \mathbb{H}_t(\epsilon) \subset \mathbb{H}_t(x)$$
$$z_t \in (\mathbb{H}_t(x))^n$$
$$\mathbb{H}_t(z) \subset \mathbb{H}_t(x)$$
$$z_t \perp \mathbb{H}_t(\epsilon).$$

The processes (z_t) and (ϵ_t) are orthogonal to each other, i.e.

$$z_t \perp \epsilon_s \text{ for all } s, t \in \mathbb{Z}.$$

For $t \geq s$, this assumption follows from $z_t \perp \mathbb{H}_t(\epsilon)$, $\epsilon_s \in (\mathbb{H}_t(\epsilon))^n$ and for $s > t$ from $\epsilon_s \perp \mathbb{H}_t(x)$, $z_t \in (\mathbb{H}_t(x))^n$. Due to the fact that $\mathbb{H}_t(y) \subset \mathbb{H}_t(\epsilon)$, the processes (y_t) and (z_t) are also orthogonal to each other. It holds that $x_t = y_t + z_t$ and hence $\mathbb{H}_t(x) \subset \mathbb{H}_t(y) \oplus \mathbb{H}_t(z)$. Then again, we also have $\mathbb{H}_t(y) \oplus \mathbb{H}_t(z) \subset \mathbb{H}_t(x)$ due to $\mathbb{H}_t(y) \subset \mathbb{H}_t(x)$ and $\mathbb{H}_t(z) \subset \mathbb{H}_t(x)$. Thus, the closed subspace $\mathbb{H}_t(x)$ is also the sum of two orthogonal closed subspaces

$$\mathbb{H}_t(x) = \mathbb{H}_t(y) \oplus \mathbb{H}_t(z).$$

Due to the fact that $\mathbb{H}_t(\epsilon) \subset \mathbb{H}_t(x) = \mathbb{H}_t(y) \oplus \mathbb{H}_t(z)$, $\mathbb{H}_t(\epsilon) \perp \mathbb{H}_t(z)$ and $\mathbb{H}_t(y) \subset \mathbb{H}_t(\epsilon)$ hold, one obtains

$$\mathbb{H}_t(\epsilon) = \mathbb{H}_t(y).$$

We now consider the prediction of y_{t+h} from its own infinite past. For this purpose, we decompose y_{t+h} into

$$y_{t+h} = \underbrace{\sum_{j=0}^{h-1} b_j \epsilon_{t+h-j}}_{\perp \mathbb{H}_t(\epsilon) = \mathbb{H}_t(y)} + \underbrace{\sum_{j \geq h} b_j \epsilon_{t+h-j}}_{\in (\mathbb{H}_t(\epsilon))^n = (\mathbb{H}_t(y))^n} .$$

Every component of the second part of the right side is contained in $\mathbb{H}_t(y) = \mathbb{H}_t(\epsilon)$ and the components of the first one are orthogonal to this space. Hence, we have the prediction

$$\hat{y}_{t,h} = P_{\mathbb{H}_t(y)}\, y_{t+h} = \sum_{j \geq h} b_j \epsilon_{t+h-j} \tag{2.19}$$

and we see that the process (y_t) is regular because

$$\underset{h \to \infty}{\text{l.i.m}}\, \hat{y}_{t,h} = \underset{h \to \infty}{\text{l.i.m}} \sum_{j \geq h} b_j \epsilon_{t+h-j} = 0.$$

The error of the one-step ahead forecast for y_{t+1} is $b_0 \epsilon_{t+1} = \epsilon_{t+1}$. This means that the ϵ_t's are the innovations of (y_t) as well.

Due to the orthogonality relation $\epsilon_t \perp \mathbb{H}_{t-1}(x)$, see (2.16) and (2.15), we can also decompose the space $\mathbb{H}_t(x)$ into a sum of orthogonal spaces as follows:

$$\begin{aligned}
\mathbb{H}_t(x) &= \text{sp}\{\epsilon_t\} \oplus \mathbb{H}_{t-1}(x) \\
&= \text{sp}\{\epsilon_t\} \oplus \text{sp}\{\epsilon_{t-1}\} \oplus \mathbb{H}_{t-2}(x) \\
&\ \ \vdots \\
&= \text{sp}\{\epsilon_t\} \oplus \cdots \oplus \text{sp}\{\epsilon_{t+1-k}\} \oplus \mathbb{H}_{t-k}(x).
\end{aligned}$$

The random vector z_t is orthogonal to $\mathbb{H}_t(\epsilon)$, i.e. to all $\epsilon_s, s \leq t$. We can thus conclude from the decomposition of $\mathbb{H}_t(x)$ above (together with $z_t \in (\mathbb{H}_t(x))^n$) that

$$z_t \in (\mathbb{H}_s(x))^n \text{ for all } s \leq t.$$

In particular, $z_{i,t+1} \in \mathbb{H}_t(x) = \mathbb{H}_t(y) \oplus \mathbb{H}_t(z)$ for $i = 1, \ldots, n$ holds and due to $z_{i,t+1} \perp \mathbb{H}_t(y)$, $z_{i,t+1} \in \mathbb{H}_t(z)$ holds as well. However, this means that

$$\hat{z}_{t,1} = P_{\mathbb{H}_t(z)}\, z_{t+1} = z_{t+1}$$

and thus it has been shown that (z_t) is a singular process.

All that remains to be proved now is the uniqueness of this Wold decomposition. Assume that $x_t = y_t + z_t$ is an (arbitrary) decomposition satisfying conditions (a)-(c). Therefore, we have $\mathbb{H}_t(x) = \mathbb{H}_t(y) \oplus \mathbb{H}_t(z)$, $\mathbb{H}_t(y) \perp \mathbb{H}_t(z)$ and hence

$$\begin{aligned}
\underset{s \to -\infty}{\text{l.i.m}} P_{\mathbb{H}_s(x)}\, x_t &= \underset{s \to -\infty}{\text{l.i.m}} (P_{\mathbb{H}_s(y)} + P_{\mathbb{H}_s(z)})(y_t + z_t) \\
&= \underbrace{\text{l.i.m}\, P_{\mathbb{H}_s(y)}\, y_t}_{=0} + \underbrace{\text{l.i.m}\, P_{\mathbb{H}_s(z)}\, z_t}_{=z_t} \\
&= z_t.
\end{aligned}$$

This means that z_t (and consequently, of course, y_t) is unique. $\square$

The proof also shows that the h-step ahead prediction for x_{t+h} (from the infinite past) is given by

$$\hat{x}_{t,h} = \hat{y}_{t,h} + z_{t+h}.$$

The following corollary is a direct consequence of the above proposition or its proof, respectively.

Corollary 2.14 *A stationary process* (x_t) *is regular if and only if it has a causal* $MA(\infty)$ *representation* $x_t = \sum_{j\geq 0} b_j \epsilon_{t-j}$ *with* $b_0 = I$ *and* $\mathbb{H}_t(x) = \mathbb{H}_t(\epsilon)$. *The* ϵ_t's *are innovations of the process* (x_t). *The following holds for the h-step ahead prediction :*

$$\hat{x}_{t,h} = \sum_{j\geq h} b_j \epsilon_{t+h-j}$$

$$u_{t,h} = \sum_{j=0}^{h-1} b_j \epsilon_{t+h-j}$$

$$\Sigma_h = \sum_{j=0}^{h-1} b_j \Sigma b'_j.$$

A consequence of Proposition 2.13 is that the prediction for x_{t+h} (from the infinite past) is obtained by predicting the regular and singular parts separately from their respective pasts and by adding then these predictions. Given that (z_t) is a harmonic process, its prediction can be based on formula (1.20). The harmonic part can be extracted by regression on harmonic functions. To predict the regular part, one first needs coefficients b_j in the Wold decomposition. Details of how to obtain the Wold representation (2.14) and hence the predictor from the second moments of (y_t) are described in relatively general terms in Rozanov (1967, Chap. 2). The most important cases for practical purposes, AR processes or ARMA processes and processes modeled by state-space systems, are discussed in the corresponding Chaps. 5 and 6 as well as 7.

The best (generally non-linear) least squares prediction is the conditional expectation $\mathbf{E}[x_{t+h} \mid x_t, x_{t-1}, \ldots]$. We now consider a regular process (x_t) as in the above conclusion, but in addition assume that the innovations (ϵ_t) are a *martingale difference sequence*, that is $\mathbf{E}[\epsilon_{t+h} \mid \epsilon_t, \epsilon_{t-1}, \ldots] = 0$ holds for all t and $h > 0$. The σ algebras generated by $\{x_s \mid s \leq t\}$ and $\{\epsilon_s \mid s \leq t\}$ are equal because of $\mathbb{H}_t(x) = \mathbb{H}_t(\epsilon)$, and thus, it follows that

$$\mathbf{E}[x_{t+h} \mid x_t, x_{t-1}, \ldots] = \hat{x}_{t,h} + \mathbf{E}[u_{t,h} \mid x_t, x_{t-1}, \ldots] =$$

$$= \hat{x}_{t,h} + \sum_{j=0}^{h-1} b_j \underbrace{\mathbf{E}[\epsilon_{t+h-1} \mid \epsilon_t, \epsilon_{t-1}, \ldots]}_{=0} = \hat{x}_{t,h}.$$

Hence, the conditional expectation is linear here, and thus, the restriction to linear functions does not imply a loss of forecasting quality. In particular, the innovation process is a martingale difference if (ϵ_t) is an IID (independent, identical and distributed) process or even stronger, if (ϵ_t) or (x_t), respectively, is a Gaussian process.

Exercise 2.15 (*Characterization of MA(q) processes*) Show that a process (x_t) with $\gamma(q) \neq 0$ and $\gamma(k) = 0$ for $|k| > q > 0$ is an MA(q) process. Note: Show first $\hat{x}_{t,h} = 0$ for $h > q$ and then use the Wold decomposition of (x_t) and the corresponding representation of the $(q + 1)$-step ahead prediction $\hat{x}_{t,q+1}$.

Exercise 2.16 Suppose $(\epsilon_t) \sim \mathrm{WN}(\sigma^2)$ is scalar white noise, z a square integrable random variable with $\mathbf{E}z\epsilon_t = 0 \,\forall t \in \mathbb{Z}$ and $(y_t = \epsilon_t - b\epsilon_{t-1})$, $b \in \mathbb{R}$ an MA(1) process. Show that

(1) the process (y_t) is regular and the process $(z_t = z)$ is singular.
(2) the process $(x_t = y_t + z)$ is neither singular nor regular. The process (y_t) is the regular part and $(z_t = z)$ is the singular part of (x_t). See (1) in Proposition 2.13.

In Sect. 4.5, we will show that the representation $y_t = \epsilon_t - b\epsilon_{t-1}$ corresponds to the Wold representation (2.14) if and only if $|b| \leq 1$. This is the only case where the ϵ_t's are also innovations of (y_t) and (x_t).

Exercise 2.17 Prove that a process (x_t) is regular if and only if the intersection of the past spaces $\mathbb{H}_s(x)$ for all $s \leq t$ only contains the zero element, i.e. $\cap_{s \leq t} \mathbb{H}_s(x) = \{0\}$.

References

E.J. Hannan, *Multiple Time Series* (Wiley, New York, 1970)

A.N. Kolmogorov, Stationary sequences in hilbert space. Bull. Moskov. Gos. Univ. Mat., **2**, 1–40. in Russian; reprinted, Selected works of A. N. Kolmogorov, Vol. 2: Theory of probability and mathematical statistics. Nauka, Moskva **1986**, 215–255 (1941)

Y.A. Rozanov, *Stationary Random Processes* (Holden-Day, San Francisco, 1967)

H. Wold, *A Study in the Analysis of Stationary Time Series*, 2nd edn. (Almqvist and Wiksell, Uppsala, 1954)

Spectral Representation

3

In this chapter, we will show that every stationary process can be approximated by a sum of harmonic oscillations with random and uncorrelated amplitudes. This means that we can approximate the process (pointwise in t) with arbitrary accuracy by a sequence of harmonic process

$$x_t \approx a_0 + a_M(-1)^t + \sum_{m=1}^{M-1} [a_m \cos(\lambda_m t) + b_m \sin(\lambda_m t)]$$

(see also (1.24)). The limit of these sums leads to an integral representation, the so-called spectral representation of stationary processes. This spectral representation is a generalization of the Fourier representation of deterministic sequences to stationary processes. This is of key importance for the theory of stationary processes as well as for their interpretation. The spectral representation defines a bijective isometry between the time domain[1] $\mathbb{H}^{\mathbb{C}}(x)$ and the so-called frequency domain $\mathbb{H}_F(x)$ of the process. Linear dynamic transformations of stationary processes (see Chap. 4) can be performed and interpreted in this frequency domain in a particularly simple manner. The Fourier representation of the covariance function corresponding to the spectral representation of the process permits an equivalent, and nice to interpret, description of the linear dependence structure of the process.

The spectral representation of the covariance function goes back to Khinchin, Wold and Cramér. The spectral representation of stationary processes can be derived

---

[1] Here, in this chapter, it is often easier to work with complex-valued random variables and hence we frequently consider the Hilbert space of complex-valued, square integrable random variables and the corresponding "complex time domain" $\mathbb{H}^C(x)$ of a stationary process; see the definitions of the "real" and of the "complex" time domain on Sects. 1.3 and 3.2, respectively.

© Springer Nature Switzerland AG 2022
M. Deistler and W. Scherrer, *Time Series Models*, Lecture Notes in Statistics 224,
https://doi.org/10.1007/978-3-031-13213-1_3

in different ways, e.g. via the spectral representation of the forward shift operator U. This was the approach used by Kolmogorov (1941). The approach here is based on Cramér (1951) and Doob (1953), who treated the univariate case. The multivariate case was first completely solved by Rosenberg (1964) and Rozanov (1967). In this approach, the spectral representation of the covariance function is derived first. The frequency domain of the process, which is isomorphic (as a Hilbert space) to the time domain, is then constructed on this basis, and the spectral representation of the process is obtained in the last step.

In this section, we show the essential steps of the proof of the spectral representation of stationary processes. Despite the fact that the structure of the proof provides essential insights, readers may skip technical details when reading it for the first time. For some technical details concerning measure theory, we also refer to the relevant literature. Only centered processes ($\mathbf{E}x_t = 0$) are considered in this chapter, unless explicitly mentioned. However, all results can be applied to the general case without much difficulty.

3.1   The Fourier Representation of the Covariance Function

In this first section, we derive the Fourier representation (spectral representation) of the covariance function pertaining to a stationary process and introduce the spectral distribution function and spectral density of a stationary process.

We use the following notation for this purpose: For a set B, $\mathbb{1}_B$ denotes the corresponding indicator function. The left-hand limit of a function $\lambda \mapsto F(\lambda)$ at point λ is denoted (if it exists) by $F(\lambda-) = \lim_{\epsilon \downarrow 0} F(\lambda - \epsilon)$.

First, we need the notion of a positive semidefinite symmetric distribution function on $[-\pi, \pi]$ and some important properties of such distribution functions. (See also Rosenberg 1964.)

Definition 3.1 (*positive semidefinite distribution function*) A function

$$F: [-\pi, \pi] \longrightarrow \mathbb{C}^{n \times n}$$

is called a *positive semidefinite distribution function* if it satisfies the following conditions:

1. $F(-\pi) = 0 \in \mathbb{C}^{n \times n}$.
2. F is monotonically nondecreasing,[2] i.e. $F(\lambda_1) \leq F(\lambda_2)$ for $\lambda_1 \leq \lambda_2$.
3. F is right-continuous.
4. (Symmetry) $F(-\lambda)' = F(\pi-) - F(\lambda-)$ for $-\pi < \lambda < \pi$ and $F(\pi) \in \mathbb{R}^{n \times n}$.

---

[2] Reminder: For two Hermitian matrices A, B, notation $A \geq B$ implies that $(A - B)$ is positive semidefinite.

It follows from conditions (1) and (2) that $0 \leq F(\lambda) \leq F(\pi)$. Thus, $F(\lambda)$ is always positive semidefinite and hence also Hermitian ($F(\lambda) = F(\lambda)^*$). The diagonal elements $F_{ii}(\cdot)$ are distribution functions of positive measures and the off-diagonal elements $F_{ij}(\cdot)$, $i \neq j$ are distribution functions of complex-valued measures. For a scalar (measurable) function $a \colon [-\pi, \pi] \to \mathbb{C}$, we interpret the integral $\int a(\lambda)dF(\lambda)$ simply componentwise, that is

$$\int_{-\pi}^{\pi} a(\lambda)dF(\lambda) = \left(\int_{-\pi}^{\pi} a(\lambda)dF_{ij}(\lambda) \right)_{i,j=1,\ldots,n} .$$

The distribution function F defines a matrix-valued measure μ_F on the Borel sigma algebra $\mathcal{B}$ on the interval $[-\pi, \pi]$:

$$B \in \mathcal{B} \longmapsto \mu_F(B) = \int_{-\pi}^{\pi} \mathbb{1}_B(\lambda)dF(\lambda) \in \mathbb{C}^{n \times n}.$$

In general, a function μ from a sigma algebra $\mathcal{B}$ to a Banach space $\mathbb{M}$ is called *measure*,[3] if $\mu(\varnothing) = 0 \in \mathbb{M}$ and μ is σ-additive, that is, for a mutually disjoint sequence $(B_j)_{j>0} \in \mathcal{B}$, the following holds:

$$\mu \left(\bigcup_{j>0} B_j \right) = \sum_{j>0} \mu(B_j).$$

The measure μ_F is called positive semidefinite, because $\mu_F(B)$ is always positive semidefinite: For each vector $a \in \mathbb{C}^{1 \times n}$, $F_{aa}(\lambda) := aF(\lambda)a^*$ is the distribution function of a positive measure and thus we have

$$a\mu_F(B)a^* = a \left[\int \mathbb{1}_B(\lambda)dF(\lambda) \right] a^* = \int \mathbb{1}_B(\lambda)dF_{aa}(\lambda) \geq 0.$$

The trace $F^\tau(\lambda) = \sum_{i=1}^{n} F_{ii}(\lambda)$ is the distribution function of a positive measure μ^τ and the measures μ_{ij} assigned to F_{ij} are absolutely continuous with respect to μ^τ: If $\mu^\tau(B) = 0$ holds, then $(\mu_{ij}(B) = 0 \in \mathbb{C}^{n \times n})$ follows from $\mu_F(B) \geq 0$. Let $f^\tau = (f_{ij}^\tau)$ be the matrix of the Radon–Nikodym derivatives[4] of μ_{ij} with respect to μ^τ. It can be easily shown that $f^\tau(\lambda) \geq 0$ and that $\operatorname{tr}(f^\tau(\lambda)) = 1$ holds μ^τ-almost everywhere. For (row-) functions $a \colon [-\pi, \pi] \to \mathbb{C}^{1 \times n}$ and $b \colon [-\pi, \pi] \to \mathbb{C}^{1 \times n}$, we now define

$$\int_{-\pi}^{\pi} \left[a(\lambda)dF(\lambda)b(\lambda)^* \right] := \int_{-\pi}^{\pi} \left[a(\lambda)f^\tau(\lambda)b(\lambda)^* \right] dF^\tau(\lambda) \qquad (3.1)$$

---

[3] μ is also called vector-valued measure, see, e.g. van Dulst (2001).
[4] See, e.g. Brokate and Kersting (2011, Radon–Nikodym theorem (IX.1)).

if the integral on the right side exists. If the integrals $\int a_i(\lambda)\overline{b_j(\lambda)}dF_{ij}(\lambda)$ exist for all $i, j = 1, \ldots, n$ (e.g. if the functions a and b are bounded), then it holds that

$$\int_{-\pi}^{\pi} \left[a(\lambda)dF(\lambda)b(\lambda)^*\right] = \sum_{i,j=1}^{n} \int_{-\pi}^{\pi} a_i(\lambda)\overline{b_j(\lambda)}dF_{ij}(\lambda).$$

Proposition 3.2 (Herglotz) *A sequence* $(\gamma(k) \in \mathbb{R}^{n \times n} \,|\, k \in \mathbb{Z})$ *is the covariance function of a stationary process if and only if a positive semidefinite distribution function* $F : [-\pi, \pi] \longrightarrow \mathbb{C}^{n \times n}$ *exists so that*

$$\gamma(k) = \int_{-\pi}^{\pi} e^{i\lambda k}dF(\lambda) \ \forall k \in \mathbb{Z}. \tag{3.2}$$

The distribution function F *is uniquely determined for given* γ.

Definition 3.3 (*spectral distribution function*) If (γ) is the covariance function of the stationary process (x_t), then the distribution function (F) defined by (3.2) is called the *spectral distribution function* of the process (x_t).

Proof (*Proof of Proposition* 3.2) First we illustrate that the sequence γ defined by (3.2) is real-valued and positive semidefinite if F is a positive semidefinite distribution function. According to Proposition 1.6, γ is hence the covariance function of a stationary process.

From the symmetry condition (4), it follows that

$$F(\pi-) = \lim_{\lambda \downarrow -\pi} \left(F(-\lambda)' + F(\lambda-)\right) = F(\pi-)'.$$

Thus, the contribution $e^{i\pi k}(F(\pi) - F(\pi-))$ of a point mass in π to the integral (3.2) is real-valued. Therefore, we can assume without loss of generality for the proof of $\gamma(k) \in \mathbb{R}^{n \times n}$ that $F(\pi) = F(\pi-)$ holds. Let $\tilde{\mu}_F$ be the pushforward measure of μ_F under the mapping $\lambda \mapsto -\lambda$, which means that $\tilde{\mu}_F(-B) = \mu_F(B)$ holds for $B \in \mathcal{B}$ and $-B := \{\lambda \,|\, -\lambda \in B\}$. For the corresponding distribution function, which we denote as $\tilde{F}$, it holds that

$$\tilde{F}(\lambda) = F(\pi) - F((-\lambda)-) = F(\lambda)' = \overline{F(\lambda)} \text{ for } -\pi < \lambda \text{ and}$$
$$\tilde{F}(-\pi) = F(\pi) - F(\pi-) = 0.$$

Using the transformation theorem for integrals, we now get

$$\gamma(k) = \int_{-\pi}^{\pi} e^{i\lambda k}dF(\lambda) = \int_{-\pi}^{\pi} e^{-i\lambda k}d\tilde{F}(\lambda) = \overline{\int_{-\pi}^{\pi} e^{i\lambda k}dF(\lambda)} = \overline{\gamma(k)}$$

which means that $\gamma(k)$ is a real matrix as claimed.

A sequence $(\gamma(k) \mid k \in \mathbb{Z})$ is positive semidefinite if and only if

$$\sum_{k,l=0}^{p-1} a_k \gamma(l-k) a_l^* \geq 0, \ \forall p \in \mathbb{N} \text{ and } a_k \in \mathbb{C}^{1 \times n}, \ k = 0, \ldots, p-1.$$

Again, this condition follows from the spectral representation (3.2) of $\gamma(k)$ by way of elementary transformations:

$$\sum_{k,l=0}^{p-1} a_k \gamma(l-k) a_l^* = \sum_{k,l=0}^{p-1} a_k \left[\int_{-\pi}^{\pi} e^{i(l-k)\lambda} dF(\lambda) \right] a_l^*$$

$$= \sum_{k,l=0}^{p-1} \int_{-\pi}^{\pi} \left[(a_k e^{-i\lambda k}) dF(\lambda) (a_l e^{-i\lambda l})^* \right]$$

$$= \int_{-\pi}^{\pi} \left[\left(\sum_{k=0}^{p-1} a_k e^{-i\lambda k} \right) f^\tau(\lambda) \left(\sum_{k=0}^{p-1} a_k e^{-i\lambda k} \right)^* \right] dF^\tau(\lambda) \geq 0.$$

Now, in order to prove the converse statement, we commence from a covariance function γ. For $q \in \mathbb{N}$, we consider the time-discrete Fourier transform of the sequence $(\gamma^{(q)}(k) \mid k \in \mathbb{Z})$, which is defined by

$$\gamma^{(q)}(k) = \begin{cases} \left(1 - \frac{|k|}{q}\right) \gamma(k) & \text{for } |k| < q \\ 0 & \text{otherwise.} \end{cases}$$

From this, it follows that

$$f^{(q)}(\lambda) = \frac{1}{2\pi} \sum_{k=-\infty}^{\infty} e^{-i\lambda k} \gamma^{(q)}(k) = \frac{1}{2\pi} \sum_{k=-q+1}^{q-1} e^{-i\lambda k} \left(1 - \frac{|k|}{q}\right) \gamma(k), \ \lambda \in [-\pi, \pi].$$

The Fourier transform $f^{(q)}(\lambda)$ is positive semidefinite, since (for $a \in \mathbb{C}^{1 \times n}$)

$$2\pi q a f^{(q)}(\lambda) a^* = \sum_{k=1-q}^{q-1} e^{-i\lambda k} (q - |k|) a \gamma(k) a^* = \sum_{k,l=0}^{q-1} (ae^{-i\lambda k}) \gamma(l-k)(ae^{-i\lambda l})^* \geq 0$$

and symmetric, i.e. $f^{(q)}(\lambda)' = f^{(q)}(-\lambda)$. This symmetry directly follows from the symmetry of the covariance function $\gamma(k)' = \gamma(-k)$. Thus, it is now easy to show that

$$F^{(q)}(\lambda) = \int_{-\pi}^{\lambda} f^{(q)}(v) dv$$

is a positive semidefinite, symmetric distribution function. The corresponding measure is denoted by $\mu_F^{(q)}$. The sequence $\gamma^{(q)}$ can be obtained from $f^{(q)}$ by the inverse Fourier transform, that is, for $k < |q|$, we have

$$\left(1 - \frac{|k|}{q}\right)\gamma(k) = \int_{-\pi}^{\pi} e^{i\lambda k} f^{(q)}(\lambda)d\lambda = \int_{-\pi}^{\pi} e^{i\lambda k} dF^{(q)}(\lambda) = \int_{-\pi}^{\pi} e^{i\lambda k} \mu_F^{(q)}(d\lambda).$$

(3.3)

$\gamma(0) = \int_{-\pi}^{\pi} \mu_F^{(q)}(d\lambda) = \mu_F^{(q)}([-\pi, \pi])$ holds in particular. The measures $\mu_F^{(q)}$ are bounded $(\mu_F^{(q)}(\Delta) \leq \gamma(0))$. Thus, it follows from the generalization of Helly's selection theorem for positive semidefinite measures (See, e.g. Fritzsche and Kirstein 1988) that a subsequence $\mu_F^{(q_r)}$ exists that converges weakly to a positive semidefinite measure μ_F. By weak convergence $\mu_F^{(q_r)} \longrightarrow \mu_F$ and (3.3), one obtains the following representation of the covariance function:

$$\gamma(k) = \lim_{r\to\infty} \left(1 - \frac{|k|}{q_r}\right)\gamma(k) = \lim_{r\to\infty} \int_{-\pi}^{\pi} e^{i\lambda k} \mu_F^{(q_r)}(d\lambda) = \int_{-\pi}^{\pi} e^{i\lambda k} \mu_F(d\lambda).$$

(3.4)

We now define a distribution function from μ_F by

$$F(\lambda) = \mu_F((-\pi, \lambda]) \text{ for } \lambda < \pi$$
$$F(\pi) = \mu_F([-\pi, \pi]).$$

This shifts a possibly existing point mass of μ_F at $-\pi$ to π in order to achieve $F(-\pi) = 0$. This operation does not change the integral (3.4), that is

$$\gamma(k) = \int_{-\pi}^{\pi} e^{i\lambda k} \mu_F(d\lambda) = \int_{-\pi}^{\pi} e^{i\lambda k} dF(\lambda) \ \forall k \in \mathbb{Z}.$$

We still have to show that the limiting distribution F constructed this way satisfies the symmetry conditions (4). Note that $F(\pi) = \mu_F([-\pi, \pi]) = \gamma(0) \in \mathbb{R}^{n\times n}$ holds. If F is continuous at points $-\lambda_1, -\lambda_2, \lambda_1, \lambda_2, -\pi < \lambda_1 < \lambda_2 < \pi$, then it follows from the symmetry of the $F^{(q)}$'s and the weak convergence $\mu_F^{(q_r)} \to \mu_F$ that

$$F(-\lambda_1)' - F(-\lambda_2)' = F(\lambda_2) - F(\lambda_1).$$

The limit for $\lambda_2 \uparrow \pi$ and $\lambda_1 \uparrow \lambda$ hence yields

$$F(-\lambda)' - \underbrace{F(-\pi)'}_{=0} = F(\pi-) - F(\lambda-).$$

The distribution function F is unique since it follows from $\int e^{i\lambda k}(dF(\lambda) - d\tilde{F}(\lambda)) = 0, \forall k \in \mathbb{Z}$ that the two positive semidefinite distribution functions are identical, i.e. $F(\lambda) = \tilde{F}(\lambda) \ \forall \lambda \in [-\pi, \pi]$. Here, one uses the fact that the trigonometric polynomials are dense in the Hilbert space of functions that are square integrable with respect to F, as will be shown at the end of Sect. 3.2. $\square$

If the spectral distribution function F of (x_t) is absolutely continuous with respect to the Lebesgue measure μ, then there exists a function $f : [-\pi, \pi] \to \mathbb{C}^{n \times n}$, such that

$$F(\lambda) = \int_{-\pi}^{\lambda} f(\mu) d\mu \ \forall \lambda \in [-\pi, \pi]. \tag{3.5}$$

It is clear that the spectral density f is only μ-a.e. uniquely determined. Of course, it holds that

$$\gamma(k) = \int_{-\pi}^{\pi} e^{i\lambda k} f(\lambda) d\lambda \ \forall k \in \mathbb{Z}. \tag{3.6}$$

Definition 3.4 (*spectral density*) The function f is called *spectral density* of the process.

The Herglotz theorem immediately leads to the following characterization of spectral densities.

Proposition 3.5 *A function* $f : [-\pi, \pi] \longrightarrow \mathbb{C}^{n \times n}$ *is the spectral density of a stationary process if and only if*

1. *f is integrable (componentwise) with respect to the Lebesgue measure μ.*
2. *$f(\lambda) \geq 0$ holds μ-a.e.*
3. *$f(-\lambda) = f(\lambda)'$ holds μ-a.e.*

The spectral density is also defined by condition (3.6) (μ-a.e.). That is, if a function f satisfies these equations, then f is a spectral density of the associated process. Based on this observation, we now obtain the following:

Corollary 3.6 *If the covariance function γ is absolutely summable $(\sum_k \|\gamma(k)\| < \infty)$, then the time-discrete Fourier transform of γ*

$$f(\lambda) = \frac{1}{2\pi} \sum_{k=-\infty}^{\infty} e^{-i\lambda k} \gamma(k), \ \lambda \in [-\pi, \pi] \tag{3.7}$$

is a spectral density of the process.

The absolute summability of the covariance function is a sufficient, but not a necessary condition for the existence of a spectral density. In case that the autocovariance function is absolutely summable, the partial sums $\frac{1}{2\pi} \sum_{k=-q}^{q} e^{-i\lambda k} \gamma(k)$ converge uniformly on $[-\pi, \pi]$ to f and the spectral density f in this case is hence continuous[5] (see, e.g. Tretter 2013, Proposition VIII.35).

Exercise 3.7 Consider a function $g: [-\pi, \pi] \longrightarrow \mathbb{R}$, $\lambda \longmapsto g(\lambda) = g_0 + 2g_1 \cos(\lambda)$. Show that g is a spectral density if and only if $g_0 \geq 0$ and $2|g_1| \leq g_0$ holds. See also Exercise 1.20 concerning the autocovariance function of MA(1) processes.

Exercise 3.8 Let $\gamma : \mathbb{Z} \longrightarrow \mathbb{R}$ be a (scalar) autocovariance function. We now consider the truncated function γ^q:

$$\gamma^q(k) = \begin{cases} \gamma(k) & \text{for } |k| \leq q \\ 0 & \text{for } |k| > q. \end{cases}$$

Find an example of γ, such that γ^q is not an autocovariance function.

Of course, not every stationary process has a spectral density. The spectral distribution function of a harmonic process, see (1.26), is a step function with finitely many jumps and therefore not absolutely continuous. For this reason, such processes do not have a spectral density.

Consider a stacked process $(z_t = (x_t', y_t')'$ and partition its spectral density (if it exists) accordingly as

$$f_z = \begin{pmatrix} f_x & f_{xy} \\ f_{yx} & f_y \end{pmatrix}$$

then f_x, f_y are called autospectra of the processes (x_t) and (y_t), respectively, and f_{yx}, f_{xy} are so-called cross-spectra between (y_t) and (x_t), respectively, as well as between (x_t) and (y_t).

For an n-dimensional process (x_t) with spectral density $f = (f_{ij})_{i,j=1,...,n}$, f_{ii} is the autospectrum of the component process (x_{it}) and f_{ij} is the cross spectrum between the (scalar) processes (x_{it}) and (x_{jt}).

3.2   The Frequency Domain of Stationary Processes

In this section, we introduce the frequency domain, a Hilbert space which is isometrically isomorphic to the time domain. Certain operations are easier to perform in the frequency domain.

---

[5] As has been noted above, the spectral density of a process is not unique, there exists an equivalence class of μ-a.e. identical spectral densities. In the case of an absolute summable autocovariance function, this class contains a unique, continuous representative.

We now consider (row) functions $a \colon [-\pi, \pi] \longrightarrow \mathbb{C}^{1 \times n}$. We say that a is square integrable with respect to F, if

$$\int_{-\pi}^{\pi} \left[a(\lambda) dF(\lambda) a(\lambda)^* \right] = \int_{-\pi}^{\pi} a(\lambda) f^{\tau}(\lambda) a(\lambda)^* dF^{\tau}(\lambda) < \infty. \qquad (3.8)$$

Now we want to endow the set of these square integrable functions—more precisely, the set of suitable equivalence classes of such functions—with a Hilbert space structure. Two functions a, b are equivalent if

$$\int_{-\pi}^{\pi} \left[(a(\lambda) - b(\lambda)) dF(\lambda) (a(\lambda) - b(\lambda))^* \right] = 0$$

holds.

If all components a_i are square integrable with respect to F^{τ}, then a is square integrable with respect to F. However, the inverse does not hold in general, as shown by the following exercise:

Exercise 3.9 Show that function a defined by $a(\lambda) = (1 - e^{-i\lambda})^{-1}(1, -1)$ for $\lambda \neq 0$ and $a(0) = (0, 0)$ is square integrable with respect to

$$F(\lambda) = \begin{pmatrix} 1 & 1 \\ 1 & 1 \end{pmatrix} (\lambda + \pi).$$

However, the components a_1, a_2 are not square integrable with respect to $F^{\tau} = 2(\lambda + \pi)$.

For positive semidefinite matrices $M \in \mathbb{C}^{n \times n}$, $M \geq 0$ one can construct a unique Hermitian, positive semidefinite root, that is a matrix $N \in \mathbb{C}^{n \times n}$ that satisfies $N \geq 0$, $N = N^*$ and $M = NN$. The root associated with M is often denoted by $M^{1/2}$ and the matrix representation of the projection onto the row space of M is denoted by M^P. Now let $f^{\tau/2}(\lambda)$ be the root of $f^{\tau}(\lambda)$ and $f^P(\lambda)$ be the corresponding projection matrix. The root $M^{1/2}$ and the projection matrix M^P are continuous functions of the elements of M. Thus, $f^{\tau/2}$ and f^P are measurable functions. We can now write the above quadratic form as

$$\int_{-\pi}^{\pi} \left[a(\lambda) dF(\lambda) a(\lambda)^* \right] = \int_{-\pi}^{\pi} \left[a(\lambda) f^{\tau}(\lambda) a(\lambda)^* \right] dF^{\tau}(\lambda)$$

$$= \int_{-\pi}^{\pi} a(\lambda) f^{\tau/2}(\lambda) \left[a(\lambda) f^{\tau/2}(\lambda) \right]^* dF^{\tau}(\lambda).$$

This means that a is square integrable with respect to F if and only if the components $(a(\cdot) f^{\tau/2}(\cdot))_k$ of $(a(\cdot) f^{\tau/2}(\cdot))$ are square integrable with respect to F^{τ}. Now if a, b

are two such square integrable functions, it then follows from this observation that the integral

$$\int_{-\pi}^{\pi} \left[a(\lambda) dF(\lambda) b(\lambda)^* \right] = \int_{-\pi}^{\pi} \left[a(\lambda) f^\tau(\lambda) b(\lambda)^* \right] dF^\tau(\lambda)$$

$$= \int_{-\pi}^{\pi} a(\lambda) f^{\tau/2}(\lambda) \left[b(\lambda) f^{\tau/2}(\lambda) \right]^* dF^\tau(\lambda)$$

exists. Thus, arbitrary linear combinations $(\alpha a(\lambda) + \beta b(\lambda))$, $\alpha, \beta \in \mathbb{C}$ are square integrable as well. We say that two square integrable functions a, b are equivalent (with respect to F), if the difference satisfies

$$\int_{-\pi}^{\pi} \left[(a(\lambda) - b(\lambda)) dF(\lambda) (a(\lambda) - b(\lambda))^* \right] = 0.$$

For the sake of simplicity, we shall use the same symbol for functions and equivalence classes in the following. The mapping

$$(a, b) \mapsto \langle a, b \rangle_F := \int_{-\pi}^{\pi} \left[a(\lambda) dF(\lambda) b(\lambda)^* \right]$$

is an inner product and the corresponding norm is denoted by $(\|a\|_F = \sqrt{\langle a, a \rangle_F})$. We denote the (complex) vector space of (equivalence classes of) square integrable row functions with the inner product defined above as $\mathbb{L}_2^{\mathbb{C}}([-\pi, \pi], \mathcal{B}, F)$. This space is complete and hence a Hilbert space.

The completeness of $\mathbb{L}_2^{\mathbb{C}}([-\pi, \pi], \mathcal{B}, F)$ can be shown in the following way: Suppose $(a_r)_{r>0}$ is a Cauchy sequence, i.e.

$$\lim_{r,s \to \infty} \int \left[(a_r - a_s) f^{\tau/2} \left[(a_r - a_s) f^{\tau/2} \right]^* \right] dF^\tau(\lambda) =$$

$$\lim_{r,s \to \infty} \sum_k \int \left| (a_r f^{\tau/2})_k - (a_s f^{\tau/2})_k \right|^2 dF^\tau(\lambda) = 0.$$

Since $\mathbb{L}_2^{\mathbb{C}}([-\pi, \pi], \mathcal{B}, F^\tau)$ is a Hilbert space, it follows that $(a_r f^{\tau/2})_k \to \tilde{a}_k \in \mathbb{L}_2^{\mathbb{C}}([-\pi, \pi], \mathcal{B}, F^\tau)$ for $k = 1, \dots, n$. We set $\tilde{a} = (\tilde{a}_1, \dots, \tilde{a}_n)$ and $a = \tilde{a} f^{\dagger/2}$, where $f^{\dagger/2}$ denotes the Moore–Penrose Inverse of $f^{\tau/2}$. With this, we obtain $a f^{\tau/2} = \tilde{a} f^p$ and because of

$$(a_r f^{\tau/2} - a f^{\tau/2})(a_r f^{\tau/2} - a f^{\tau/2})^* = (a_r f^{\tau/2} - \tilde{a}) f^p (a_r f^{\tau/2} - \tilde{a})^*$$

$$\leq (a_r f^{\tau/2} - \tilde{a})(a_r f^{\tau/2} - \tilde{a})^*,$$

we get the desired convergence of $a_n \longrightarrow a$ (with respect to $\| \cdot \|_F$). See Rozanov (1967, Chap. I, Lemma 7.1) and Rosenberg (1964).

Exercise 3.10 Note that the size of the equivalence classes depends on the distribution function F. For instance, if F is a step function with jumps at points λ_k, where $F(\lambda_k) - F(\lambda_k-) > 0$ holds, then two functions are equivalent if and only if they take the same values at the jump points. On the other hand, if there exists a spectral density, which is positive definite for all frequencies, then two functions are equivalent if and only if they coincide μ-a.e. Prove these claims.

We could choose another (scalar) distribution function F^* such that the elements of F are absolutely continuous with respect to F^* and then perform the same construction with this distribution function instead of F^τ. However, as shown by Rozanov (1967, p. 29f), both the integral (3.8) and the Hilbert space $\mathbb{L}_2^{\mathbb{C}}([-\pi, \pi], \mathcal{B}, F)$ do not depend on the particular choice of the distribution function F^*. For instance, if F is absolutely continuous with respect to the Lebesgue measure μ, i.e. if a spectral density f exists, then we have

$$\int_{-\pi}^{\pi} \left[a(\lambda) dF(\lambda) b(\lambda)^* \right] = \int_{-\pi}^{\pi} a(\lambda) f(\lambda) b(\lambda)^* d\lambda.$$

Definition 3.11 (*Frequency domain*) Assuming that F is the spectral distribution function of the process (x_t), then $\mathbb{H}_F(x) := \mathbb{L}_2^{\mathbb{C}}([-\pi, \pi], \mathcal{B}, F)$ is called the *frequency domain* of (x_t).

Exercise 3.12 Show that for a measurable, bounded function $a: [-\pi, \pi] \to \mathbb{C}, a \in \mathbb{H}_F(x)$ holds for any scalar, stationary process (x_t). To be more precise, there is an equivalence class in $\mathbb{H}_F$ containing a.

In the following, we often consider instead of the real Hilbert space $\mathbb{L}_2(\Omega, \mathcal{A}, \mathbf{P})$ the Hilbert space of complex-valued, square integrable random variables $\mathbb{L}_2^{\mathbb{C}}(\Omega, \mathcal{A}, \mathbf{P})$ with the complex numbers as multipliers. The inner product is defined by $\langle x, y \rangle = \mathbf{E} x \bar{y}$. Accordingly, the complex time domain $\mathbb{H}^{\mathbb{C}}(x)$ of a stationary process (x_t) is the subspace generated by the components x_{it} in $\mathbb{L}_2^{\mathbb{C}}(\Omega, \mathcal{A}, \mathbf{P})$:

$$\mathbb{H}^{\mathbb{C}}(x) = \overline{\text{sp}}\{x_{it}, i = 1, \ldots, n, \ t \in \mathbb{Z}\} \subset \mathbb{L}_2^{\mathbb{C}}(\Omega, \mathcal{A}, \mathbf{P}).$$

Clearly, we can consider $\mathbb{H}(x)$ to be a subset of $\mathbb{H}^{\mathbb{C}}(x)$. Although we consider only real processes, the complex Hilbert space $\mathbb{H}^{\mathbb{C}}(x)$ often allows for a simpler representation. Moreover, the transition to the complex time domain does not lead to "undesirable complex results", as can be seen in the following exercise, for example:

Exercise 3.13 Let $y \in \mathbb{H}(x)$ and $\mathbb{M} \subset \mathbb{H}(x)$ be a closed subspace of $\mathbb{H}(x)$. Suppose the subspace generated by $\mathbb{M}$ in $\mathbb{H}^{\mathbb{C}}(x)$ is $\mathbb{M}^{\mathbb{C}} = \overline{\text{sp}}\{x \mid x \in \mathbb{M}\} \subset \mathbb{H}^{\mathbb{C}}(x)$. Now show that the projection of y onto $\mathbb{M}$ (in $\mathbb{H}(x)$) yields the same result as the projection of y onto $\mathbb{M}^{\mathbb{C}}$ (in $\mathbb{H}^{\mathbb{C}}(x)$).

In the following, u_k denotes the kth (row) unit vector in $\mathbb{C}^{1 \times n}$. Furthermore, $e^{i \cdot t}$ denotes the function $\lambda \mapsto e^{i \lambda t}$.

Proposition 3.14 *The mapping*

$$x_{kt} \in \mathbb{H}^{\mathbb{C}}(x) \longmapsto u_k e^{i \cdot t} \in \mathbb{H}_F(x), \quad k = 1, \dots, n, \ t \in \mathbb{Z}$$

can be uniquely extended to a bijective isometry

$$\Phi \colon \mathbb{H}^{\mathbb{C}}(x) \longrightarrow \mathbb{H}_F(x)$$

between the time domain $\mathbb{H}^{\mathbb{C}}(x)$ and the frequency domain $\mathbb{H}_F(x)$.

As an isometry between two Hilbert spaces, the mapping Φ is linear and continuous.

Proof According to the construction of the spectral measure F, it holds that

$$\langle x_{kt}, x_{ls} \rangle = \gamma_{kl}(t - s) = u_k \left[\int e^{i \lambda (t-s)} d F(\lambda) \right] u_l^*$$

$$= \int_{-\pi}^{\pi} \left[(u_k e^{i \lambda t}) d F(\lambda) (u_l e^{i \lambda s})^* \right] = \langle u_k e^{i \cdot t}, u_l e^{i \cdot s} \rangle_F.$$

It is easy to see that this isometry property is also preserved for the linear extension and finally for the continuous extension (see also the discussion on the forward shift in the section on the time domain of stationary processes). This proves that Φ is a bijective isometry between the complex time domain and the Hilbert subspace generated by the trigonometric polynomials in $\mathbb{H}_F(x) = \mathbb{L}_2^{\mathbb{C}}([-\pi, \pi], \mathcal{B}, F)$. Thus, in order to complete the proof of the proposition, we have to show that the trigonometric polynomials are dense in $\mathbb{L}_2^{\mathbb{C}}([-\pi, \pi], \mathcal{B}, F)$, i.e.

$$\overline{\mathrm{sp}}\{u_k e^{i \cdot t} \mid k = 1, \dots, n, \ t \in \mathbb{Z}\} = \mathbb{L}_2^{\mathbb{C}}([-\pi, \pi], \mathcal{B}, F).$$

Rozanov (1967) argues here as follows: The scalar trigonometric polynomials are dense in $\mathbb{L}_2^{\mathbb{C}}([-\pi, \pi], \mathcal{B}, F_{kk})$ and functions $a = (a_1, \dots, a_n)$ with $a_k \in \mathbb{L}_2^{\mathbb{C}}([-\pi, \pi], \mathcal{B}, F_{kk})$ are dense in $\mathbb{L}_2^{\mathbb{C}}([-\pi, \pi], \mathcal{B}, F)$. $\qquad \square$

Exercise 3.15 Show

$$\Phi(\mathbb{H}(x)) = \{a \in \mathbb{H}_F(x) \mid a(\lambda) = \bar{a}(-\lambda) \ \mu^{\tau} - \text{a.e.}\}$$

3.3 Spectral Representation of Stationary Processes

Using the inverse mapping Φ^{-1}, we can assign to each function $a \in \mathbb{H}_F(x)$ an element in $\mathbb{H}^{\mathbb{C}}(x)$. This, in particular, holds for the functions of the form $u_k \mathbb{1}_B$, where $B \in \mathcal{B}$. Hence, we can assign to each Borel set B a random vector

$$z(B) = (z_k(B))_{k=1,\dots,n}, \quad z_k(B) = \Phi^{-1}(u_k \mathbb{1}_B) \in \mathbb{H}^{\mathbb{C}}(x).$$

This mapping is σ-additive because Φ^{-1} is linear and continuous: For a sequence of disjoint Borel sets $(B_m)_{m>0}$, it holds that

$$z_k\left(\bigcup_{m=1}^{\infty} B_m\right) = \Phi^{-1}\left(\sum_{m=1}^{\infty} u_k \mathbb{1}_{B_m}\right) = \sum_{m=1}^{\infty} \Phi^{-1}(u_k \mathbb{1}_{B_m}) = \sum_{m=1}^{\infty} z_k(B_m).$$

Thus, z is a "random measure," or more precisely, a random variable-valued measure

$$z \colon \mathcal{B} \longrightarrow (\mathbb{L}_2^{\mathbb{C}})^n.$$

See also Rozanov (1967). For two Borel sets B_1, $B_2 \in \mathcal{B}$, it follows that

$$\begin{aligned}
\mathbf{E}z_k(B_1)\overline{z_l(B_2)} &= \left\langle \Phi^{-1}(u_k \mathbb{1}_{B_1}), \Phi^{-1}(u_l \mathbb{1}_{B_2}) \right\rangle = \left\langle u_k \mathbb{1}_{B_1}, u_l \mathbb{1}_{B_2} \right\rangle_F \\
&= \int_{-\pi}^{\pi} \left[\mathbb{1}_{B_1}(\lambda) u_k dF(\lambda) \mathbb{1}_{B_2}(\lambda) u_l^* \right] = \mu_{kl}(B_1 \cap B_2) \\
\mathbf{E}z(B_1)z(B_2)^* &= \mu_F(B_1 \cap B_2).
\end{aligned} \tag{3.9}$$

Using

$$(z(\lambda) = z([-\pi, \lambda]) \mid \lambda \in [-\pi, \pi]),$$

we now define the so-called *spectral process* of process (x_t). This spectral process can be interpreted as a "random distribution function". Our notation is a bit sloppy, since we use the same symbol for random measure $z(B)$ and the random distribution function $z(\lambda)$. The respective meaning of $z(.)$ should be obvious from the context.

Proposition 3.16 *The spectral process has the following properties:*

1. $z(-\pi) = 0$ a.s.
2. $\mathbf{E}z(\lambda)^* z(\lambda) < \infty$ for all $\lambda \in [-\pi, \pi]$ (the process is square integrable).
3. $\mathbf{E}z(\lambda) = 0$ (the expectations are zero).
4. $\text{l.i.m.}_{\epsilon \downarrow 0} z(\lambda + \epsilon) = z(\lambda)$ for $\lambda \in [-\pi, \pi)$ (the process is right-continuous).
5. $\mathbf{E}\left[(z(\lambda_4) - z(\lambda_3))(z(\lambda_2) - z(\lambda_1))^* \right] = 0$ for $-\pi \le \lambda_1 < \lambda_2 \le \lambda_3 < \lambda_4 \le \pi$ (the increments of the process are orthogonal).
6. $\mathbf{E}\left[z(\lambda)z(\lambda)^* \right] = F(\lambda)$ for $-\pi \le \lambda \le \pi$.
7. $\mathbf{E}\left[(z(\lambda_2) - z(\lambda_1))(z(\lambda_2) - z(\lambda_1))^* \right] = F(\lambda_2) - F(\lambda_1)$ for $-\pi \le \lambda_1 < \lambda_2 \le \pi$.

The properties (2) and (3) follow from $z_k(\lambda) \in \mathbb{H}^{\mathbb{C}}(x)$ and $\mathbf{E}x_t = 0$. All other properties are direct consequences of (3.9) and the continuity on the right of F.

Definition 3.17 (*Process with orthogonal increments*) A stochastic process $(z(\lambda) \mid \lambda \in [-\pi, \pi])$ with complex-valued random vectors $z(\lambda): \Omega \to \mathbb{C}^n$ is called *process with orthogonal increments* if the above properties (1)–(5) are satisfied.

Using the spectral process, we can now give an "explicit" representation of the inverse $\Phi^{-1}: \mathbb{H}_F(x) \longrightarrow \mathbb{H}^{\mathbb{C}}(x)$. First we consider simple functions $a \in \mathbb{H}_F(x)$, i.e. functions of the form

$$
a(\lambda) = \sum_{m=1}^{M} a_m \mathbb{1}_{B_m}(\lambda), \quad a_m = (a_{m1}, \ldots, a_{mn}) \in \mathbb{C}^{1 \times n}, \ B_m \in \mathcal{B}.
$$

Due to the linearity of Φ^{-1}, it is immediately clear that we have

$$
\Phi^{-1}(a) = \Phi^{-1}\left(\sum_{m=1}^{M} \left(\sum_{k=1}^{n} a_{mk}(u_k \mathbb{1}_{B_m}) \right) \right) = \sum_{m=1}^{M} a_m z(B_m).
$$

Every function $a \in \mathbb{H}_F(x)$ can be approximated by using simple functions. This means that a sequence of simple functions $(a^{(k)} \in \mathbb{H}_F(x))_{k \geq 1}$ exists, so that $a = \lim_k a^{(k)}$. Due to the continuity of Φ^{-1}, we thus get

$$
\Phi^{-1}(a) = \underset{k}{\text{l.i.m}} \ \Phi^{-1}(a^{(k)}).
$$

Due to this construction, the inverse mapping Φ^{-1} is given by the stochastic integral with respect to z:

$$
\Phi^{-1}(a) = \int_{-\pi}^{\pi} a(\lambda) dz(\lambda).
$$

As we have already seen, it is often more elegant to work with random vectors whose components are elements of the time domain $\mathbb{H}^{\mathbb{C}}(x)$. Analogously, we often consider matrix functions $a: [-\pi, \pi] \longrightarrow \mathbb{C}^{m \times n}$ as well, whose rows a_k, $k = 1, \ldots, m$ are elements of the frequency domain $\mathbb{H}_F(x)$. We shall now compile some useful definitions and notations for such random vectors and matrix functions, respectively.

For random vectors $y = (y_1, \ldots, y_m)'$, $y \in (\mathbb{H}^{\mathbb{C}}(x))^m$, means that $y_k \in \mathbb{H}^{\mathbb{C}}(x)$ for $k = 1, \ldots, m$. For a matrix function $a: [-\pi, \pi] \longrightarrow \mathbb{C}^{m \times n}$, we write $a \in (\mathbb{H}_F(x))^m$ if all rows a_k, $k = 1, \ldots, m$ are elements of the frequency domain $\mathbb{H}_F(x)$. For random vectors $y \in (\mathbb{H}^{\mathbb{C}}(x))^m$ and matrices $a \in (\mathbb{H}_F(x))^m$, one defines $\Phi(y)$ resp. $\Phi^{-1}(a)$ simply componentwise or row-wise, respectively. That is,

$$\Phi(y) := (\Phi(y_1)', \ldots, \Phi(y_m)')' \text{ and}$$

$$\Phi^{-1}(a) = \int_{-\pi}^{\pi} a(\lambda)dz(\lambda) := \begin{pmatrix} \Phi^{-1}(a_1) \\ \vdots \\ \Phi^{-1}(a_m) \end{pmatrix} = \begin{pmatrix} \int_{-\pi}^{\pi} a_1(\lambda)dz(\lambda) \\ \vdots \\ \int_{-\pi}^{\pi} a_m(\lambda)dz(\lambda) \end{pmatrix}.$$

For the case $a(\lambda) = a_0(\lambda)I_n$, where $a_0 : [-\pi, \pi] \longrightarrow \mathbb{C}$ is a scalar, complex-valued function, we can also use the short notation $\Phi^{-1}(a) = \int a(\lambda)dz(\lambda) = \int a_0(\lambda)dz(\lambda)$. The "inner product" of two random vectors $y \in (\mathbb{H}^{\mathbb{C}}(x))^m$, $z \in (\mathbb{H}^{\mathbb{C}}(x))^n$ is the matrix

$$\langle y, z \rangle := (\langle y_k, z_l \rangle)_{k,l} = \mathbf{E}yz^*$$

and for $a \in (\mathbb{H}_F(x))^m$, $b \in (\mathbb{H}_F(x))^n$, we write

$$\langle a, b \rangle_F := (\langle a_k, b_l \rangle_F)_{k,l} = \int_{-\pi}^{\pi} \left[a(\lambda)dF(\lambda)b(\lambda)^* \right] = \int_{-\pi}^{\pi} \left[a(\lambda)f^{\tau}(\lambda)b(\lambda)^* \right] dF^{\tau}(\lambda).$$

Using this notation and the isometry of Φ^{-1}, we now obtain

$$\mathbf{E}\left[\left(\int a(\lambda)dz(\lambda) \right) \left(\int b(\lambda)dz(\lambda) \right)^* \right] = \langle \Phi^{-1}(a), \Phi^{-1}(b) \rangle = \langle a, b \rangle_F \quad (3.10)$$

$$= \int_{-\pi}^{\pi} \left[a(\lambda)dF(\lambda)b(\lambda)^* \right] \quad (3.11)$$

and for the important case that a spectral density f exists, we get

$$\mathbf{E}\left[\left(\int a(\lambda)dz(\lambda) \right) \left(\int b(\lambda)dz(\lambda) \right)^* \right] = \int_{-\pi}^{\pi} \left[a(\lambda)f(\lambda)b(\lambda)^* \right] d\lambda. \quad (3.12)$$

The integral representation $x_t = \Phi^{-1}(e^{i \cdot t}I_n)$ is of particular importance. This representation

$$x_t = \int_{-\pi}^{\pi} e^{i\lambda t}dz(\lambda) \quad (3.13)$$

is called *spectral representation* of the stationary process (x_t). The harmonic function $e^{i \cdot t}$ can be (uniformly) approximated by a concrete sequence of step functions:

$$e^{i\lambda t} \approx \sum_{k=0}^{K-1} e^{i\lambda_k^K t} \mathbb{1}_{(\lambda_k^K, \lambda_{k+1}^K]}(\lambda), \quad \lambda_k^K = -\pi + \frac{2\pi k}{K}, \quad k = 0, \ldots, K.$$

Hence, we obtain an approximation of the process (x_t) by a sequence of harmonic processes as follows:

$$x_t = \int_{-\pi}^{\pi} e^{i\lambda t}dz(\lambda) = \lim_{K \to \infty} \sum_{k=0}^{K-1} e^{i\lambda_k^K t}(z(\lambda_{k+1}^K) - z(\lambda_k^K)) \ \forall t \in \mathbb{Z}.$$

The above spectral representation (3.13) is (essentially) unique, which means that if $(\tilde{z}(\lambda) \mid \lambda \in [-\pi, \pi])$ is a process with orthogonal increments and if the process (x_t) can be represented as

$$x_t = \underset{K \to \infty}{\text{l.i.m}} \sum_{k=0}^{K-1} e^{i\lambda_k^K t}(\tilde{z}(\lambda_{k+1}^K) - \tilde{z}(\lambda_k^K)) \; \forall t \in \mathbb{Z},$$

then $z(\lambda) = \tilde{z}(\lambda) \, a.s.$ holds for all $\lambda \in [-\pi, \pi]$. Here, we shall present only an outline of how to prove the uniqueness of the spectral representation. A process $\tilde{z}$ with orthogonal increments defines a random measure on $[-\pi, \pi]$ and hence one can construct the stochastic integral (with respect to $\tilde{z}$) quite analogously to above. The integral $\int a \, d\tilde{z}(\lambda)$ is defined for functions $a \in \mathbb{L}_2([-\pi, \pi], \mathcal{B}, \tilde{F})$, where $\tilde{F} = \mathbf{E}\tilde{z}(\lambda)\tilde{z}(\lambda)^*$ is a positive semidefinite distribution function. Then, in particular, we get $\gamma(k) = \mathbf{E}x_k x_0^* = \mathbf{E}(\int e^{i\lambda k} d\tilde{z}(\lambda))(\int e^{i\lambda 0} d\tilde{z}(\lambda))^* = \int_{-\pi}^{\pi} e^{i\lambda k} d\tilde{F}(\lambda)$. Due to the fact that the spectral distribution function is unique, it follows that $\tilde{F} = F$ and $\mathbb{L}_2^{\mathbb{C}}([-\pi, \pi], \mathcal{B}, \tilde{F}) = \mathbb{L}_2^{\mathbb{C}}([-\pi, \pi], \mathcal{B}, F) = \mathbb{H}_F(x)$. Now $x_t = \int e^{i\lambda t} dz(\lambda) = \int e^{i\lambda t} d\tilde{z}(\lambda)$ holds for all $t \in \mathbb{Z}$ and since the trigonometric polynomials are dense in $\mathbb{L}_2^{\mathbb{C}}([-\pi, \pi], \mathcal{B}, F)$, $\int a(\lambda) dz(\lambda) = \int a(\lambda) d\tilde{z}(\lambda)$ holds for all $a \in \mathbb{L}_2^{\mathbb{C}}([-\pi, \pi], \mathcal{B}, F)$. For the indicator function $u_k \mathbb{1}_{[-\pi, \lambda]}$, it now follows that

$$z_k(\lambda) = \int u_k \mathbb{1}_{[-\pi, \lambda]}(\nu) dz(\nu) = \int u_k \mathbb{1}_{[-\pi, \lambda]}(\nu) d\tilde{z}(\nu) = \tilde{z}_k(\lambda).$$

The above equation is an identity in $\mathbb{L}_2^{\mathbb{C}}(\Omega, \mathcal{A}, \mathbf{P})$, thus only $z(\lambda) = \tilde{z}(\lambda)$ a.s. holds.

We summarize these results in the following proposition:

Proposition 3.18 (Spectral representation of stationary processes) *For every stationary (centered) process (x_t), there exists a process with orthogonal increments $(z(\lambda) \mid \lambda \in [-\pi, \pi])$ such that*

$$x_t = \int_{-\pi}^{\pi} e^{i\lambda t} dz(\lambda) = \underset{K \to \infty}{\text{l.i.m}} \sum_{k=0}^{K-1} e^{i\lambda_k^K t}(z(\lambda_{k+1}^K) - z(\lambda_k^K)) \; \forall t \in \mathbb{Z}. \qquad (3.14)$$

In addition, the spectral process $(z(\lambda) \mid \lambda \in [-\pi, \pi])$ is unique (almost everywhere) and $z_j(\lambda) \in \mathbb{H}^{\mathbb{C}}(x)$, for $j = 1, \ldots, n$.

This result shows that every stationary process can be approximated with arbitrary accuracy by a harmonic process $(x_t^K = \sum_{k=0}^{K-1} e^{i\lambda_k^K t}(z(\lambda_{k+1}^K) - z(\lambda_k^K)))$. However, this approximation is not uniform in t, which means that in general, $\sup_t \mathbf{E}(x_t - x_t^K)'(x_t - x_t^K)$ with $K \to \infty$ does not converge to zero. This also explains the seeming contradiction that one can describe a regular process as a "limit" of singular processes.

Exercise 3.19 We consider a process $(z(\lambda) \mid \lambda \in [-\pi, \pi])$ with $z(\lambda) = \sigma W(\lambda + \pi)$, where $(W(s) \mid s \in \mathbb{R}, s \geq 0)$ is a Wiener process (Brownian motion). Show that z is a process with orthogonal increments and that the process $x_t = \int_{-\pi}^{\pi} e^{i\lambda t} dz(\lambda)$ generated by z is Gaussian white noise.

The spectral representation can also be defined for processes with expectation $\mathbf{E}x_t \neq 0$. In this case, the spectral process $z(\cdot)$ has a jump $z(0) - z(0-)$ at the frequency $\lambda = 0$ and it holds that $\mathbf{E}(z(0) - z(0-)) = \mathbf{E}x_0 = \mathbf{E}x_t$. The spectral distribution F here also has a point of discontinuity at $\lambda = 0$ and hence the spectral density does not exist. Moreover in this case, F corresponds to the non-centered second moments $\mathbf{E}x_s x_0' = \gamma(s) + \mathbf{E}x_0(\mathbf{E}x_0)'$ rather than to the autocovariance function $\gamma(\cdot)$.

Note that $\mathrm{U}x_t = x_{t+1} = \int_{-\pi}^{\pi} e^{i\lambda(t+1)} dz(\lambda) = \int_{-\pi}^{\pi} e^{i\lambda} e^{i\lambda t} dz(\lambda)$ and thus the application of the forward shift in the time domain corresponds to multiplication by the function $e^{i\cdot}$ in the frequency domain, which means that the diagram below commutes

$$
\begin{array}{ccc}
\mathbb{H}^{\mathbb{C}}(x) & \xrightarrow{\;\mathrm{U}^k\;} & \mathbb{H}^{\mathbb{C}}(x) \\[4pt]
\Phi \downarrow & & \uparrow \Phi^{-1} \\[4pt]
\mathbb{H}_F(x) & \xrightarrow{\;(e^{i\cdot k})\cdot\;} & \mathbb{H}_F(x)
\end{array}
$$

It is therefore simpler to analyze filters in the frequency domain, as we shall see in the next chapter.

Interpretation of the Spectrum

The spectral representation (3.14) represents the process (x_t) approximatively as a sum of harmonic oscillations. Now we consider a frequency band $B = (\lambda_1, \lambda_2] \subset (0, \pi)$ and we want to quantify more precisely the contribution of the associated oscillations to the process. To decompose the process into *real* components, we also need to consider the "mirrored" frequency band $-B = [-\lambda_2, -\lambda_1)$. We define the two indicator functions

$$
\mathbb{1}_1 = \mathbb{1}_{B \cup -B} \text{ and } \mathbb{1}_2 = 1 - \mathbb{1}_1
$$

and decompose x_t accordingly into two components

$$
x_t = \int_{-\pi}^{\pi} e^{i\lambda t} dz(\lambda) = \underbrace{\int_{-\pi}^{\pi} \mathbb{1}_1(\lambda) e^{i\lambda t} dz(\lambda)}_{=:x_t^{(1)}} + \underbrace{\int_{-\pi}^{\pi} \mathbb{1}_2(\lambda) e^{i\lambda t} dz(\lambda)}_{=:x_t^{(2)}} .
$$

The first component $x_t^{(1)}$ is the part of the process generated by oscillations with frequencies in the band $(\lambda_1, \lambda_2]$ and $x_t^{(2)}$ is the "remainder". The two components are orthogonal to each other because the corresponding frequency domains do not overlap ($\mathbb{1}_1 \mathbb{1}_2 = 0$). The following holds (see Eq. (3.10) and the symmetry conditions for F):

$$\mathbf{E}x_t^{(1)}(x_t^{(2)})' = \int_{-\pi}^{\pi} \mathbb{1}_1(\lambda)e^{i\lambda t}\overline{(\mathbb{1}_2(\lambda)e^{i\lambda t})}dF(\lambda) = 0,$$

$$\mathbf{E}x_t^{(1)}(x_t^{(1)})' = \int_{-\pi}^{\pi} \underbrace{|\mathbb{1}_1(\lambda)e^{i\lambda t}|^2}_{=\mathbb{1}_1(\lambda)} dF(\lambda)$$

$$= \underbrace{(F(\lambda_2) - F(\lambda_1))}_{:=\Delta F} + \underbrace{(F((-\lambda_1)-) - F((-\lambda_2)-))}_{=(\Delta F)'} = \Delta F + (\Delta F)',$$

$$\mathbf{E}x_t^{(2)}(x_t^{(2)})' = \int_{-\pi}^{\pi} \underbrace{|\mathbb{1}_2(\lambda)e^{i\lambda t}|^2}_{=\mathbb{1}_2(\lambda)} dF(\lambda),$$

$$\gamma(0) = \mathbf{E}x_t x_t' = \int_{-\pi}^{\pi} (\mathbb{1}_1(\lambda) + \mathbb{1}_2(\lambda))dF(\lambda) = \mathbf{E}x_t^{(1)}(x_t^{(1)})' + \mathbf{E}x_t^{(2)}(x_t^{(2)})'.$$

The quotient

$$0 \le \frac{2\Delta F_{kk}}{\gamma_{kk}(0)} \le 1$$

is the relative part of the variance (the latter is often interpretable as power in electrical engineering) of the k-th component x_{kt}, which is explained by the oscillations with frequencies in interval $(\lambda_1, \lambda_2]$. Frequency intervals in which the *(auto-)spectral distribution* F_{kk} grows relatively strongly (that is, the increment ΔF_{kk} is relatively large) are hence "important" for the process. If the spectral density exists, then we can also write this increment as

$$\Delta F = \int_{\lambda_1}^{\lambda_2} f(\nu)d\nu \approx f(\lambda_1)(\lambda_2 - \lambda_1).$$

Of course, the approximation $\Delta F \approx f(\lambda_1)(\lambda_2 - \lambda_1)$ is the better the smaller $(\lambda_2 - \lambda_1)$ and the smoother f is. Peaks in the *spectral density* f_{kk} thus show "important frequencies" (resp. frequency bands) (Fig. 3.1).

$(\Delta F_{kl} + \overline{\Delta F_{kl}})$ is the covariance between $x_{kt}^{(1)}$ and $x_{lt}^{(1)}$, i.e. a measure of the linear dependency between x_{kt} and x_{lt} in the frequency band $(\lambda_1, \lambda_2]$. For small intervals, it holds again that $\Delta F_{kl} \approx f_{kl}(\lambda_1)(\lambda_2 - \lambda_1)$ and we can hence interpret the cross-spectral density $f_{kl}(\lambda)$ as a measure of the linear dependence of the two components x_{kt} and x_{lt} in the "vicinity" of frequency λ. We shall discuss this in more detail in Sect. 4.4 on the Wiener filter.

Frequencies $\lambda = 0$ and $\lambda = \pi$ must be discussed separately. However, the above said can also be applied analogously to these frequencies.

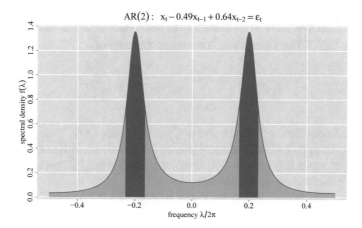

Fig. 3.1 This figure shows the spectral density of an AR(2) process $x_t = 0.49x_{t-1} - 0.64x_{t-2} + \epsilon_t$. The frequency band highlighted in dark blue explains 50% of the power of the process

Examples

Harmonic processes

The spectral process of a scalar, real-valued, stationary, harmonic process $x_t = \sum_k z_k e^{i\lambda_k t}$ (see (1.25) and (1.26)) is

$$z(\lambda) = \sum_{\{k \mid \lambda_k \leq \lambda\}} z_k$$

and the spectral distribution function is

$$F(\lambda) = \mathbf{E}|z(\lambda)|^2 = \sum_{\{k \mid \lambda_k \leq \lambda\}} \mathbf{E}|z_k|^2.$$

The distribution function F is a monotonically nondecreasing right-continuous step function (with $F(-\pi) = 0$ and $F(\pi) = \mathbf{E}|z(\pi)|^2 = \mathbf{E}x_t^2 < \infty$) and it thus defines a *discrete* measure on the interval $[-\pi, \pi]$. Hence, the spectral density does not exist. The jump points mark the frequencies λ_k and the jump heights are a measure for the corresponding amplitudes, more precisely the jump heights are equal to the expectation of the squared absolute values of the (random) amplitudes.

White noise

Let $(\epsilon_t) \sim \text{WN}(\Sigma)$ be white noise. The spectral density of (ϵ_t)

$$f_\epsilon(\lambda) = \frac{1}{2\pi} \sum_{k=-\infty}^{\infty} \gamma_\epsilon(k) e^{-i\lambda k} = \frac{1}{2\pi} \Sigma \tag{3.15}$$

is constant (independent of the frequency). (See Corollary 3.6.) This means that all frequencies (frequency bands of equal length) are equally important for the process. In analogy to white light, which is a uniform superposition of all colors ($\doteq$ frequencies), this process is hence called "white noise".

Conversely, if the spectral density of a process (x_t) is constant ($f(\lambda) \equiv f_0$), it follows that

$$\gamma(k) = \int_{-\pi}^{\pi} f(\lambda) e^{i\lambda k} d\lambda = f_0 \int_{-\pi}^{\pi} e^{i\lambda k} d\lambda = \begin{cases} 2\pi f_0 & \text{for } k = 0 \\ 0 & \text{for } k \neq 0. \end{cases}$$

This means that a process is white noise if and only if the spectral density is constant.

MA(q) processes

Let $(x_t = b_0 \epsilon_t + \cdots + b_q \epsilon_{t-q})$ be an MA(q) process, where $(\epsilon_t) \sim \text{WN}(\Sigma)$. The spectral density of (x_t) is

$$
\begin{aligned}
f(\lambda) \quad &= \quad \frac{1}{2\pi} \sum_{k=-q}^{q} \gamma(k) e^{-i\lambda k} = \frac{1}{2\pi} \sum_{k=-q}^{q} \sum_{j=0}^{q} b_{j+k} e^{-i\lambda(j+k)} \Sigma b'_j e^{i\lambda j} \\
&\underset{(j+k)\to k}{=} \frac{1}{2\pi} \sum_{k=0}^{q} \sum_{j=0}^{q} b_k e^{-i\lambda k} \Sigma b'_j e^{i\lambda j} \\
&= \quad \left(\sum_{j=0}^{q} b_j e^{-ij\lambda} \right) \left(\frac{1}{2\pi} \Sigma \right) \left(\sum_{j=0}^{q} b_j e^{-ij\lambda} \right)^*.
\end{aligned}
\tag{3.16}
$$

Here, as in Eq. (1.8), we have extended the coefficients b_j with zeros for the sake of simplicity, which means that we set $b_j = 0$ for $j < 0$ and $j > q$.

MA(∞) processes

For MA(∞) processes $x_t = \sum_{j=-\infty}^{\infty} b_j \epsilon_{t-j}$ with absolutely summable coefficients ($\sum_j \|b_j\| < \infty$), the autocovariance function is also absolutely summable and thus (x_t) has a spectral density. In complete analogy to the MA(q) case, one obtains

$$f(\lambda) = b(\lambda) f_\epsilon(\lambda) b(\lambda)^* \tag{3.17}$$

where $f_\epsilon(\lambda) = \frac{1}{2\pi}\Sigma$ is the spectral density of the white noise (ϵ_t) and

$$b(\lambda) = \sum_{j=-\infty}^{\infty} b_j e^{-i\lambda j}.$$

We will show in the next chapter that every MA(∞) process ($x_t = \sum_j b_j \epsilon_{t-j}$, $\epsilon_t \sim WN(\Sigma)$) has a spectral density, that is

$$f(\lambda) = \frac{1}{2\pi} b(\lambda)\Sigma b(\lambda)^*,$$

where $b(\lambda) = \sum_j b_j e^{-i\lambda j}$. This infinite sum converges in $\mathbb{L}_2([-\pi, \pi], \mathcal{B}, \mu)$ due to $\sum_j \|b_j\|^2 < \infty$. (Here μ denotes the Lebesgue measure.) Of course, this holds in particular for any regular process as can be seen from the Wold representation. Conversely, the existence of a spectral density is not sufficient for the regularity of the process. In particular, a process is singular if its spectral density is equal to zero on a non-void interval. See, e.g. Brockwell and Davis (1991, Example 5.6.1) as an example. A more precise characterization yields the following result, which we state without a proof (Rozanov (1967, Propositions 6.1 and 6.2 in Chap. II)).

Proposition 3.20 (Generalized Szegő theorem) *A multivariate stationary process with spectral density f, which has full rank μ—a.e., is regular if and only if*

$$\int_{-\pi}^{\pi} \log(\det(f(\lambda)) d\lambda > -\infty.$$

In this case, it holds that

$$\det \Sigma = \exp\left(\frac{1}{2\pi} \int_{-\pi}^{\pi} \log(\det(2\pi f(\lambda)) d\lambda\right)$$

where Σ denotes the variance of the innovations of (x_t).

Finally, we suggest that readers solve the following problems.

Exercise 3.21 Let (x_t) be a scalar, centered stationary process with an autocovariance function $\gamma(k)$ and spectral density $f(\lambda)$. We now consider trigonometric polynomials of the form $a(\lambda) = \sum_{k=0}^{q} a_k e^{-i\lambda k}$, $a_k \in \mathbb{R}$. Show the following:

1. $a \in \mathbb{H}_F(x)$ and $\Phi^{-1}(ae^{i\cdot t}) = \sum_{k=0}^{q} a_k x_{t-k}$.
2. If a, b are two such polynomials, then it holds that

$$\langle ae^{i\cdot t}, be^{i\cdot t} \rangle_F = \int_{-\pi}^{\pi} a(\lambda)e^{i\lambda t} f(\lambda)\overline{b(\lambda)e^{i\lambda t}} d\lambda$$

$$= \sum_{k,l=0}^{q} a_k \gamma(l-k) b_l = (a_0, \ldots, a_q)\Gamma_{q+1}(b_0, \ldots, b_q)'$$

with $\Gamma_{q+1} = (\gamma(l-k))_{k,l=1,\ldots,q+1}$.

3.

$$\frac{1}{2\pi} \int_{-\pi}^{\pi} a(\lambda)\overline{b(\lambda)} d\lambda = \sum_{k=0}^{q} a_k b_k.$$

4. If $\sum_{k=0}^{q} a_k^2 = 1$ holds, it follows that

$$\underline{\lambda}(\Gamma_{q+1}) \leq (a_0, \ldots, a_q)\Gamma_{q+1}(a_0, \ldots, a_q)' \leq \overline{\lambda}(\Gamma_{q+1})$$

$$\inf_{\lambda \in [-\pi,\pi]} f(\lambda) \leq \quad \frac{1}{2\pi}\int_{-\pi}^{\pi} |a(\lambda)|^2 f(\lambda) d\lambda \quad \leq \sup_{\lambda \in [-\pi,\pi]} f(\lambda)$$

where $0 \leq \underline{\lambda}(\Gamma_{q+1}) \leq \overline{\lambda}(\Gamma_{q+1})$ denotes the minimum or resp. maximum eigenvalue of the Toeplitz matrix Γ_{q+1}. These inequalities imply the following relationship between the spectral density f and the eigenvalues of the Toeplitz matrices Γ_{q+1}:

$$2\pi \inf_{\lambda \in [-\pi,\pi]} f(\lambda) \leq \underline{\lambda}(\Gamma_{q+1}) \text{ and } \overline{\lambda}(\Gamma_{q+1}) \leq 2\pi \sup_{\lambda \in [-\pi,\pi]} f(\lambda).$$

Exercise 3.22 Let $a(\lambda) = \sum_{k=-\infty}^{\infty} a_k e^{-i\lambda k}$, $a_k \in \mathbb{R}$ and (x_t) be a scalar process with spectral density f. Prove the following:

1. If the spectral density f has an upper bound ($\sup_{\lambda \in [-\pi,\pi]} f(\lambda) \leq c < \infty$) and if the coefficients $(a_k \mid k \in \mathbb{Z})$ are square summable ($\sum_k a_k^2 < \infty$), then $\sum_{k=-\infty}^{\infty} a_k e^{-i\lambda k}$ exists as a limit in the $\mathbb{H}_F(x)$ sense, i. e. $a \in \mathbb{H}_F(x)$.
2. If the spectral density has a *lower* bound ($\inf_{\lambda \in [-\pi,\pi]} f(\lambda) \geq c > 0$) and if $\sum_{k=-\infty}^{\infty} a_k e^{-i\lambda k}$ exists as limit in the $\mathbb{H}_F(x)$ sense ($a \in \mathbb{H}_F(x)$), the coefficients are thus square summable ($\sum_k a_k^2 < \infty$).
3. If (x_t) is white noise, then $a \in \mathbb{H}_F(x)$ holds if and only if the coefficients are square summable.

Hint: Use the above Exercise 3.21 and compare also Proposition 1.22.

Exercise 3.23 Let (x_t) and (y_t) be two n-dimensional, centered stationary processes that are uncorrelated to each other (i.e. $\mathbf{E}(x_s y_t') = 0 \ \forall s, t \in \mathbb{Z}$). Now show the following assertions (in obvious notation) for the "sum process" $(z_t = x_t + y_t)$:

$$\gamma_z(k) = \gamma_x(k) + \gamma_y(k)$$
$$z_z(\lambda) = z_x(\lambda) + z_y(\lambda)$$
$$F_z(\lambda) = F_x(\lambda) + F_y(\lambda)$$
$$f_z(\lambda) = f_x(\lambda) + f_y(\lambda) \ \text{ if } f_x, \ f_y \text{ exist.}$$

Exercise 3.24 Verify the following representation of the covariance function γ of a process with spectral density f:

$$\gamma_{ij}(k) = 2 \int_0^\pi \left[\cos(k\lambda)\Re(f_{ij}(\lambda)) - \sin(k\lambda)\Im(f_{ij}(\lambda)) \right] d\lambda \ \text{ for } i \neq j,$$
$$\gamma_{ii}(k) = 2 \int_0^\pi \left[\cos(k\lambda)\Re(f_{ii}(\lambda)) \right] d\lambda.$$

Exercise 3.25 (*Sampling and Aliasing*) Let $(x_t \mid t \in \mathbb{Z})$ be a stationary process with the spectral density f_x and $\Delta \in \mathbb{N}$. Show that the process $(y_s = x_{\Delta s} \mid s \in \mathbb{Z})$ is stationary with the spectral density

$$f_y(\lambda) = \frac{1}{\Delta} \sum_{j=0}^{\Delta-1} f_x\left(\frac{\lambda + 2\pi j}{\Delta} \right).$$

Hint: Verify the following equation(s):

$$\gamma_y(k) = \gamma_x(k\Delta) = \int_{-\pi}^\pi e^{i\lambda\Delta k} f_x(\lambda) d\lambda = \int_{-\pi}^\pi e^{i\lambda k} f_y(\lambda) d\lambda.$$

References

P. Brockwell, R. Davis, *Time Series: Theory and Methods*, Springer Series in Statistics, 2nd edn. (Springer, New York, 1991)

M. Brokate, G. Kersting, *Maß und Integral* (Mathematik Kompakt. Springer Basel AG, Basel, 2011). 9783034606462

H. Cramér, A contribution to the theory of stochastic processes, in *Proceedings of 2nd Berkeley Symposium on Mathematical Statistics and Probabilities*, pp. 57–78 (University of California Press, 1951)

J.L. Doob, *Stochastic Processes* (Wiley, 1953)

B. Fritzsche, B. Kirstein, Schwache Konvergenz nichtnegativ hermitescher Borelmaße. Wiss. Z. Karl-Marx-Univ. Leipzig Math.-Natur. **37**(4), 375–398 (1988)

A.N. Kolmogorov, Stationary sequences in hilbert space. Bull. Moskov. Gos. Univ. Mat., **2**, 1–40. in Russian; reprinted, Selected works of A. N. Kolmogorov, Vol. 2: Theory of probability and mathematical statistics. Nauka, Moskva **1986**, 215–255 (1941)

M. Rosenberg, The square-integrability of matrix-valued functions with respect to a non-negative Hermitian measure. Duke Math. J. **31**(2), 291–298, 06 (1964). https://doi.org/10.1215/S0012-7094-64-03128-X

Y.A. Rozanov, *Stationary Random Processes* (Holden-Day, San Francisco, 1967)

C. Tretter, *Analysis I Mathematik Kompakt* (Springer Basel, Basel, 2013)

D. van Dulst, Vector measures, in *Encyclopedia of Mathematics* ed. by M. Hazewinkel (Springer, 2001). ISBN 978-1-55608-010-4

Linear, Time-Invariant, Dynamic Filters and Difference Equations

4

In this chapter, we consider linear, time-invariant and generally dynamic transformations of stationary processes. Such transformations are also called filters or systems, where the original process is the input and the transformed process is the output.

The most important application areas of such filters or systems are

- Such systems serve as (mathematical) models for real systems (e.g. technical or economic systems).
- Filters are computational methods, e.g. for the extraction ("filtering out") of components such as disturbances or seasonal fluctuations from time series.
- Linear transformations are used, for example, to build up other processes, such as $MA(\infty)$ processes, from "elementary" processes, especially white noise. Thus, the filter contains essential information about the transformed process.

Initially, we consider general linear transformations of stationary processes and then so-called l_1 filters, both in the time and the frequency domain. One section is dedicated to the interpretation of such filters in the frequency domain. The penultimate section of this chapter discusses the Wiener filter. The last section deals with the solution of linear difference equations. These solutions are obtained by so-called rational filters. They will be central to our further discussion.

© Springer Nature Switzerland AG 2022
M. Deistler and W. Scherrer, *Time Series Models*, Lecture Notes in Statistics 224,
https://doi.org/10.1007/978-3-031-13213-1_4

4.1   Linear, Time-Invariant Transformations of Stationary Processes in Time and Frequency Domain

We assign an output process (y_t) to a stationary input process (x_t) by

$$y_t = \underset{q \to \infty}{\text{l.i.m}} \sum_{j=-q}^{q} a_j^q x_{t-j}, \ a_j^q \in \mathbb{R}^{m \times n}, \ t \in \mathbb{Z}. \tag{4.1}$$

The filter (4.1) (which is understood as a mapping $(x_t) \longmapsto (y_t)$) is described by the so-called *weight function* $(a_j^q \mid j = -q, \ldots, q, \ q \in \mathbb{N})$. For the limit on the right-hand side of (4.1) to exist, the coefficients must of course satisfy (usually input-dependent) conditions. See also the discussion of $\text{MA}(\infty)$ processes, in particular Proposition 1.22. If the limit exists for one t, it exists for all $t \in \mathbb{Z}$ due to the stationarity of the input. It is immediately evident from (4.1) that this transformation is *linear* and *time-invariant*, the latter because the coefficients a_j^q do not depend on t. Here, time-invariant means that the time-shifted input process $(x_{t-s} \mid t \in \mathbb{Z})$ corresponds to the time-shifted output process $(y_{t-s} \mid t \in \mathbb{Z})$. If $a_j^q = 0 \, \forall j \neq 0$, $q \in \mathbb{N}$ holds, the filter is called *static*, otherwise the filter is *dynamic*. If $a_j^q = 0 \, \forall j < 0$, $q \in \mathbb{N}$, the filter is called *causal*.

If $\mathbb{H}(x)$ is the time domain of (x_t), it clearly holds that

$$y_{it} \in \mathbb{H}(x), \ i = 1, \ldots, m, \ t \in \mathbb{Z}.$$

Hence, the time domain $\mathbb{H}(y) = \overline{\text{sp}}\{y_{it} \mid i = 1, \ldots, m, \ t \in \mathbb{Z}\}$ is a closed subspace of $\mathbb{H}(x)$. Let $\text{U}: \mathbb{H}(x) \longrightarrow \mathbb{H}(x)$ be the forward shift associated with the process (x_t), see Proposition 1.15, i.e. $\text{U} \, x_{it} = x_{i,t+1}$ resp. $(\text{U} \, x_t = x_{t+1})$, in "vector form". From the linearity and continuity of U, it follows that

$$y_{t+1} = \underset{q \to \infty}{\text{l.i.m}} \sum_{j=-q}^{q} a_j^q x_{t+1-j} = \underset{q \to \infty}{\text{l.i.m}} \sum_{j=-q}^{q} a_j^q \, \text{U} \, x_{t-j} = \text{U} \, \underset{q \to \infty}{\text{l.i.m}} \sum_{j=-q}^{q} a_j^q x_{t-j} = \text{U} \, y_t.$$

Thus, the restriction of U to $\mathbb{H}(y)$ is the forward shift of (y_t). The unitarity of U implies directly that the "stacked" process $((x_t', y_t')' \mid t \in \mathbb{Z}) = (\text{U}^t (x_0', y_0')' \mid t \in \mathbb{Z})$ is stationary. Moreover, for the expectation $\mathbf{E} y_t$ and the covariances $\gamma_{yx}(k)$ and $\gamma_y(k)$, we obtain immediately by the bilinearity and the continuity of the inner product the following formulas:

$$\mathbf{E} y_t = \lim_{q \to \infty} \sum_{j=-q}^{q} a_j^q \mathbf{E} x_t \tag{4.2}$$

$$\gamma_{yx}(k) = \mathbf{Cov}(y_k, x_0) = \lim_{q \to \infty} \sum_{j=-q}^{q} a_j^q \gamma_x(k - j) \tag{4.3}$$

$$\gamma_y(k) = \mathbf{Cov}(y_k, y_0) = \lim_{q \to \infty} \sum_{j,l=-q}^{q} a_j^q \gamma_x(k + l - j)(a_l^q)'. \tag{4.4}$$

Hence, we have shown the following proposition.

Proposition 4.1 *If (y_t) results from (x_t) by linear transformation (4.1), $(x'_t, y'_t)'$ is stationary and the cross-covariance function $\gamma_{yx}(k) = \text{Cov}(y_{t+k}, x_t)$ is given by (4.3) and the (auto-)covariance function $\gamma_y(k) = \text{Cov}(y_{t+k}, y_t)$ is given by (4.4).*

In the following, we shall only discuss centered processes, since the expectations can be discussed quite simply using (4.2).

Exercise 4.2 Consider the MA(1) process $x_t = \epsilon_t - \epsilon_{t-1}$, $(\epsilon_t) \sim \text{WN}(\sigma^2)$. Show that

$$\underset{q \to \infty}{\text{l.i.m}} \sum_{j=0}^{q-1} \frac{q-j}{q} x_{t-j}$$

exists. Note: First show $\sum_{j=0}^{q-1} \frac{q-j}{q} x_{t-j} = \epsilon_t - \frac{1}{q} \sum_{j=1}^{q} \epsilon_{t-j}$ and then use this result to prove $\text{l.i.m}_{q \to \infty} \sum_{j=0}^{q-1} \frac{q-j}{q} x_{t-j} = \epsilon_t$.

Now assume that $(x_t) \sim \text{WN}(\sigma^2)$ is white noise. Show that in this case, the linear transformation above does *not* exist. Note: Analyze the variance of $\sum_{j=0}^{q-1} \frac{q-j}{q} x_{t-j}$ for $q \to \infty$.

Now consider the frequency domain $\mathbb{H}_F(x)$ which is isometrically isomorphic to the time domain $\mathbb{H}^{\mathbb{C}}(x)$, according to Proposition 3.14. The so-called *transfer function*

$$a(\lambda) = \lim_{q \to \infty} \sum_{j=-q}^{q} a_j^q e^{-i\lambda j} \tag{4.5}$$

is the frequency domain analogue of the weight function $(a_j^q \mid j = -q, \ldots, q, q \in \mathbb{N})$ in the time domain. The limit in (4.5) is to be understood (row-wise) with respect to the convergence in $\mathbb{H}_F(x)$. It is easy to see that the transfer function a is the image of $y_0 \in (\mathbb{H}^{\mathbb{C}}(x))^n$ under the isometry Φ, i.e. $a = \Phi(y_0)$. Moreover, it also holds that

$$y_0 = \Phi^{-1}(a) = \int_{-\pi}^{\pi} a(\lambda) dz(\lambda) \tag{4.6}$$

and

$$y_t = \Phi^{-1}(e^{i \cdot t} a) = \int_{-\pi}^{\pi} e^{i\lambda t} a(\lambda) dz(\lambda). \tag{4.7}$$

It follows from $\mathbb{H}^{\mathbb{C}}(y) \subset \mathbb{H}^{\mathbb{C}}(x)$ that $\mathbb{H}_F(y) \subset \mathbb{H}_F(x)$. The above equations show that the output of the filter (and thus the filter itself) is uniquely determined by the transfer function a.

For the sake of simplicity, we will assume below that the process (x_t) has a spectral density f_x. Then, due to the isometry between the time and the frequency domain (see Eq. (3.12)), it holds that

$$\mathbf{E} \begin{pmatrix} x_k \\ y_k \end{pmatrix} \begin{pmatrix} x_0 \\ y_0 \end{pmatrix}' = \int_{-\pi}^{\pi} e^{i\lambda k} \begin{pmatrix} I_n \\ a(\lambda) \end{pmatrix} f_x(\lambda) \begin{pmatrix} I_n \\ a(\lambda) \end{pmatrix}^* d\lambda, \quad \forall k \in \mathbb{Z}. \qquad (4.8)$$

Since (4.8) holds for all $k \in \mathbb{Z}$, the spectral density $\begin{pmatrix} f_x & f_{xy} \\ f_{yx} & f_y \end{pmatrix}$ of the "stacked" process exists and the following holds:

$$f_{yx}(\lambda) = a(\lambda) f_x(\lambda) \qquad (4.9)$$
$$f_y(\lambda) = a(\lambda) f_x(\lambda) a(\lambda)^*. \qquad (4.10)$$

Thus, we have shown the following.

Proposition 4.3 *If (x_t) has a spectral density f_x and if (y_t) was generated by a linear transformation (4.1) from (x_t), the spectral density of the process $(x_t', y_t')'$ exists and (4.9) as well as (4.10) hold.*

This proposition shows in particular that every $MA(\infty)$ process (and hence every regular process, too) has a spectral density.

It is easy to see that, in general, if the spectral densities do not necessarily exist, it holds that

$$f_{yx}^\tau(\lambda) = a(\lambda) f_x^\tau(\lambda)$$
$$f_y^\tau(\lambda) = a(\lambda) f_x^\tau(\lambda) a(\lambda)^*,$$

where $\begin{pmatrix} f_x^\tau & f_{xy}^\tau \\ f_{yx}^\tau & f_y^\tau \end{pmatrix}$ are the Radon–Nikodym derivatives of the spectral distribution function of $(x_t', y_t')'$ with respect to the "trace measure" μ^τ of (x_t).

Exercise 4.4 Let $a : [-\pi, \pi] \longrightarrow \mathbb{C}^{n \times n}$ be a unitary transfer function which means that $a(\lambda)(a(\lambda))^* = I_n$ for all $\lambda \in [-\pi, \pi]$ and let $(\epsilon_t) \sim WN(\Sigma)$ be white noise with $\Sigma = \sigma^2 I_n$, $\sigma^2 \in \mathbb{R}$. Verify that the transformed process $(x_t = \Phi^{-1}(e^{i \cdot t} a))$ is also white noise. The corresponding filter is called *all-pass filter*.

Exercise 4.5 A rational function $\underline{k}(z) = \frac{1 - \overline{z_0} z}{z - z_0}$, $z_0 \in \mathbb{C}$ is called the *Blaschke factor*. Show that

$$\underline{k}(z) \overline{\underline{k}(z)} = 1 \quad \forall z \in \mathbb{C} \text{ with } |z| = 1.$$

Blaschke factors are therefore unitary on the unit circle (i.e. $|\underline{k}(e^{i\lambda})| \equiv 1$). Calculate the product of the Blaschke factors

$$\frac{1 - \overline{z_0} z}{z - z_0} \frac{1 - z_0 z}{z - \overline{z_0}}$$

belonging to $z_0 = (a + ib)$ and $\overline{z_0} = (a - ib)$.

Exercise 4.6 Consider the MA(1) process $x_t = \epsilon_t - \epsilon_{t-1}$, $(\epsilon_t) \sim \mathrm{WN}(\sigma^2)$. Show that the transfer function

$$a(\lambda) = \frac{1}{1 - e^{-i\lambda}}$$

is an element of the frequency domain $\mathbb{H}_F(x)$. The function a is the transfer function of the filter defined in the exercise above. Using the identity

$$\sum_{j=0}^{q-1} \frac{q-j}{q} z^j = \frac{z(z^q - 1) + q(1-z)}{q(1-z)^2} \quad \text{for } z \neq 1,$$

show that $a^{(q)}(\lambda) := \sum_{j=0}^{q-1} \frac{q-j}{q} e^{-i\lambda j}$ converges pointwise for all $\lambda \neq 0$ to $a(\lambda)$ as well as in the $\mathbb{H}_F(x)$ sense. Hint:

$$\left(a^{(q)}(\lambda) - \frac{1}{1 - e^{-i\lambda}} \right) (1 - e^{-i\lambda}) = -\frac{1}{q}(e^{-i\lambda} + \cdots + e^{-i\lambda q})$$

and

$$\frac{1}{2\pi} \int_{-\pi}^{\pi} \left| \frac{1}{q}(e^{-i\lambda} + \cdots + e^{-i\lambda q}) \right|^2 d\lambda = \frac{1}{q}.$$

This argument is an exact copy of the argument in the time domain.

4.2 l_1-Filter

For a given weight function, the linear transformation (4.1) is usually not defined for every stationary input process (x_t). In the following, we consider so-called l_1-filters for which (4.1) exists for all stationary inputs:

Definition 4.7 (l_1 -filter) If the weight function of a linear transformation (4.1) is of the form

$$(a_j^q = a_j \mid j = -q, \ldots, q, \; q \in \mathbb{N}) \;\; \text{and if} \; \sum_{j=-\infty}^{\infty} \|a_j\| < \infty, \tag{4.11}$$

then the linear transformation (4.1) is called l_1 -filter.

Here any matrix norm, which is equivalent to the Frobenius norm, see the remark after Definition 1.21, may be used. We will use the spectral norm, unless the contrary is stated explicitly.

Proposition 4.8 *For any l_1-filter, the infinite sum*

$$\sum_{j=-\infty}^{\infty} a_j x_{t-j} \qquad (4.12)$$

exists for every stationary process (x_t) in the sense of the convergence in mean square.

Proof We define $\|x\| = \sqrt{\mathbf{E}x'x}$ for square integrable random vectors x. As can be easily seen, the triangle inequality $\|x + y\| \leq \|x\| + \|y\|$ holds and for a matrix with spectral norm $\|a\|$ it follows that $\|ax\| \leq \|a\|\|x\|$. The sequence of the partial sums $\sum_{j=-q}^{q} a_j x_{t-j}$ converges because

$$\|\sum_{m<|j|\leq q} a_j x_{t-j}\| \leq \|x_t\| \sum_{m<|j|\leq q} \|a_j\| \leq \|x_t\| \sum_{m<|j|} \|a_j\|$$

and because $\sum_{j=-\infty}^{\infty} \|a_j\| < \infty$ holds by assumption. $\qquad\square$

This proof shows that the convergence of (4.12) also holds more generally for input processes with bounded second moments. The weight function $(a_j \mid j \in \mathbb{Z})$ of the filter is also called *impulse response* due to the fact that an "impulse" $(x_t = \delta_{0t} u \mid t \in \mathbb{Z})$, $u \in \mathbb{R}^n$ as input yields an output $(y_t = a_t u \mid t \in \mathbb{Z})$. The Kronecker symbol δ_{st} is defined by $\delta_{st} = 1$ for $s = t$ and $\delta_{st} = 0$ otherwise.

One obtains an elegant representation of filters by introducing the *lag operator* L. For arbitrary stochastic processes, one defines

$$L(x_t \mid t \in \mathbb{Z}) := (x_{t-1} \mid t \in \mathbb{Z}).$$

The operator, which is inverse to L, exists and is denoted by L^{-1}, and L^k for $k \in \mathbb{Z}$ is defined by $L^k(x_t) = (x_{t-k})$. Of course it holds that $L^k L^r = L^{k+r}$. We can now interpret a filter with weight function $(a_j \mid j \in \mathbb{Z})$ as a formal Laurent series in the lag operator

$$\underline{a}(L) = \sum_j a_j L^j$$

and we can accordingly write

$$(y_t) = \underline{a}(L)(x_t) = (\sum_j a_j L^j)(x_t)$$

for the output (y_t) of the filter with an input (x_t). This operator notation also makes explicit that a filter is a mapping (operator) that assigns an output $(y_t = \sum_j a_j x_{t-j})$ to an input (x_t).

We distinguish the lag operator L from the backward shift U^{-1}. The backward shift U^{-1} is defined on the time domain $\mathbb{H}(x)$, i.e. on a set of random variables, while the lag operator is defined on sets of stochastic processes. The stationarity of the underlying process is essential for the construction of the backward shift (or the forward shift operator U, respectively). Note that $x_t = x_s$ and $x_{t-1} \neq x_{s-1}$ for $t \neq s$ may hold for a general process. In this case, $x_t \longmapsto x_{t-1}$ is not a well-defined mapping on $\{x_s \mid s \in \mathbb{Z}\}$. See also Proposition 1.15. On the other hand, the lag operator is well defined for arbitrary sets of stochastic processes.

The transfer function of an l_1-filter $\underline{a}(L) = \sum_{j=-\infty}^{\infty} a_j L^j$ is

$$a(\lambda) = \sum_{j=-\infty}^{\infty} a_j e^{-i\lambda j}. \tag{4.13}$$

The sum (4.13) does not only exist as a limit in $\mathbb{H}_F$, but it also exists pointwise for each $\lambda \in [-\pi, \pi]$. Moreover, this convergence is uniform on this interval and the limit function $a(\cdot)$ is continuous. The one-to-one relation between the filter coefficients (impulse response) (4.11) and the transfer function is given by the Fourier transform

$$a_j = \frac{1}{2\pi} \int_{-\pi}^{\pi} a(\lambda) e^{i\lambda j} d\lambda.$$

The transfer function $a(\cdot)$ is bounded, which implies that

$$\int_{-\pi}^{\pi} a(\lambda) f^\tau(\lambda) a^*(\lambda) dF^\tau(\lambda)$$

exists for stationary inputs (x_t) with arbitrary spectral distribution $F(\lambda)$. This shows from another point of view that for l_1-filters, the output is defined for all stationary inputs.

Exercise 4.9 Show that if $a: [-\pi, \pi] \longrightarrow \mathbb{C}^{m \times n}$ is a bounded transfer function (i.e. $\|a(\lambda)\| \leq c < \infty \ \forall \lambda \in [-\pi, \pi]$), then the output of the corresponding filter is well-defined for all stationary inputs!

Exercise 4.10 Let (x_t) be a scalar process, $\underline{k}(L) = \sum_{j=-\infty}^{\infty} k_j L^j$, $k_j \in \mathbb{R}$ an l_1-filter and $(y_t) = \underline{k}(L)(x_t)$. Prove the following inequality:

$$\gamma_y(0) \leq \gamma_x(0) \left(\sum_{j=-\infty}^{\infty} |k_j| \right)^2.$$

One often considers the series connection of filters, as shown in the following figure:

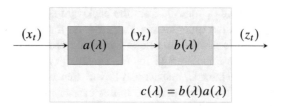

$$c(\lambda) = b(\lambda)a(\lambda)$$

The input process (x_t) is mapped by the filter with transfer function $a(\lambda)$ to the process (y_t), and this in turn is input for the filter with transfer function $b(\lambda)$.

Proposition 4.11 (Series connection of l_1-filters) *If* $\underline{a}(L) = \sum_j a_j L^j, a_j \in \mathbb{R}^{m \times n}$ *and* $\underline{b}(L) = \sum_j b_j L^j, b_j \in \mathbb{R}^{l \times m}$ *are two l_1-filters, then the series connection of these two filters is an l_1-filter* $\underline{c}(L) = \sum_j c_j L^j$ *again. That is, more precisely, if* (x_t) *is a stationary process,* $y_t = \sum_j a_j x_{t-j}$ *and* $z_t = \sum_j b_j y_{t-j}$, *then it holds that* $z_t = \sum_j c_j x_{t-j}$, *where the weight function* (c_j) *is given by*

$$c_j = \sum_{k=-\infty}^{\infty} b_k a_{j-k}, \quad \sum_j \|c_j\| < \infty \tag{4.14}$$

and the corresponding transfer function $c(\lambda)$ is equal to

$$c(\lambda) = \sum_j c_j e^{-i\lambda j} = \sum_j b_j e^{-i\lambda j} \sum_j a_j e^{-i\lambda j} = b(\lambda)a(\lambda). \tag{4.15}$$

Proof It holds that

$$z_t = \sum_j b_j y_{t-j} = \sum_j \sum_k b_j a_k x_{t-j-k}$$
$$= \sum_j \sum_k b_k a_{j-k} x_{t-j} = \sum_j c_j x_{t-j}.$$

We may rearrange the summands here, since it holds that

$$\sum_j \sum_k \|b_j a_k x_{t-j-k}\| < \infty.$$

This follows from

$$\sum_{j,k} \|b_j a_k x_{t-j-k}\| \leq \sum_{j,k} \|b_j\| \|a_k\| \|x_{t-j-k}\| = \|x_0\| \sum_j \|b_j\| \sum_k \|a_k\| < \infty.$$

The weight function (c_j) is absolutely summable, since

$$\sum_j \|c_j\| = \sum_j \left\| \sum_k b_k a_{j-k} \right\| \leq \sum_j \sum_k \|b_k\| \|a_{j-k}\| = \left(\sum_j \|b_j\| \right) \left(\sum_j \|a_j\| \right) < \infty$$

and the representation of the transfer function follows from

$$c(\lambda) = \sum_j c_j e^{-i\lambda j} = \sum_j \sum_k (b_k e^{-i\lambda k})(a_{j-k} e^{-i\lambda(j-k)}) = b(\lambda)a(\lambda).$$

$\square$

With the help of the lag operator introduced above, we can also write this result as

$$\underline{b}(L)(\underline{a}(L)(x_t)) = \underbrace{(\underline{b}(L)\underline{a}(L))}_{\underline{c}(L)}(x_t) = \underline{c}(L)(x_t).$$
$$\underbrace{\phantom{\underline{b}(L)(\underline{a}(L)(x_t))}}_{(y_t)}$$

The weight function of the series connection of the two filters is computed by convolution of the weight functions of the two filters. In this sense, the series connection of filters, i.e. Laurent series in the lag operator, corresponds to the multiplication of Laurent series (in a complex variable).

We now consider a square l_1-filter $\underline{a}(L) = \sum_j a_j L^j$, $a_j \in \mathbb{R}^{n \times n}$. An l_1-filter $\underline{b}(L) = \sum_j b_j L^j$, $b_j \in \mathbb{R}^{n \times n}$ is called the inverse filter of $\underline{a}(L)$, if the series connection of $\underline{a}(L)$ and $\underline{b}(L)$, resp. of $\underline{b}(L)$ and $\underline{a}(L)$ yields the identity, i.e. $\underline{a}(L)\underline{b}(L) = \underline{b}(L)\underline{a}(L) = I_n L^0$. For the transfer function, it thus holds that

$$b(\lambda)a(\lambda) = a(\lambda)b(\lambda) = I_n \ \forall \lambda \in [-\pi, \pi].$$

Hence, the transfer function $a(\lambda)$ must be non-singular for all $\lambda \in [-\pi, \pi]$, i.e.

$$\det a(\lambda) \neq 0 \ \forall \lambda \in [-\pi, \pi] \qquad (4.16)$$

and the transfer function of the inverse filter is equal to

$$b(\lambda) = a(\lambda)^{-1} \ \forall \lambda \in [-\pi, \pi]. \qquad (4.17)$$

It follows from Wiener's $1/f$ theorem that the condition (4.16) is also sufficient for the existence of an inverse l_1-filter; see, for example, Newman (1975). We will discuss the important special case of rational l_1-filters in more detail in Sect. 4.5.

4.3   Interpretation of Filters in the Frequency Domain

In this section, we show that linear filters are particularly simple to interpret in the frequency domain. We restrict ourselves to l_1-filters and to so-called SISO filters (single input, single output), which are filters with scalar input and output processes. We write the transfer function $k(\lambda) = \sum_{j=-\infty}^{\infty} k_j e^{-i\lambda j}$ of the l_1 filter in polar coordinate representation as $k(\lambda) = r(\lambda)e^{i\Phi(\lambda)}$. The function $r: \lambda \mapsto r(\lambda) = |k(\lambda)|$ is called *amplitude response (gain)* of the filter and $\phi: \lambda \mapsto \Phi(\lambda) = \Im(\log(k(\lambda))$ is called the *phase response* of the filter. The phase response $\phi(\lambda)$ is not determined for $k(\lambda) = 0$ and for $k(\lambda) \neq 0$, $\phi(\lambda)$ is determined only up to integer multiples of 2π. Hence, one can, for example, assume that $\phi(\lambda) \in [-\pi, \pi)$ holds. Since the filter coefficients are real, it follows that $r(-\lambda) = r(\lambda)$ and it holds for $r(\lambda) > 0$ and $-\pi < \phi(\lambda)$ that $\phi(-\lambda) = -\phi(\lambda)$. In particular, $k(\lambda) \in \mathbb{R}$ holds for $\lambda \in \{-\pi, 0, \pi\}$.

We begin the discussion with a scalar, harmonic process

$$x_t = \sum_{m=-M+1}^{M} e^{i\lambda_m t} z_m$$

$$= a_0 + a_M(-1)^t + \sum_{m=1}^{M-1} a_m \cos(\lambda_m t) + b_m \sin(\lambda_m t)$$

with frequencies $0 = \lambda_0 < \lambda_1 < \cdots < \lambda_M = \pi$ and $\lambda_{-m} = -\lambda_m$. See also (1.21) and (1.24). The output process $(y_t = \sum_j a_j x_{t-j})$ has the spectral representation (see (4.7))

$$y_t = \int_{-\pi}^{\pi} e^{i\lambda t} k(\lambda) dz(\lambda)$$

$$= \sum_{m=-M+1}^{M} e^{i\lambda_m t} k(\lambda_m) z_m = \sum_{m=-M+1}^{M} e^{i(\lambda_m t + \phi(\lambda_m))} r(\lambda_m) z_m$$

$$= k(0)a_0 + k(\pi)a_M(-1)^t +$$

$$\sum_{m=1}^{M-1} r(\lambda_m)a_m \cos(\lambda_m t + \phi(\lambda_m)) + r(\lambda_m)b_m \sin(\lambda_m t + \phi(\lambda_m)).$$

This means that the output process—just like the input process—is a linear combination of harmonic oscillations (with the same frequencies). However, the filter modifies the amplitudes by the factor $r(\lambda_m)$ and leads to a phase shift $\phi(\lambda_m)$. The oscillations with frequency λ_m are amplified or attenuated by the factor $r(\lambda_m)$, which means, in particular, that one can "filter out" certain frequencies. The phase shift $\phi(\lambda_m)$ corresponds to a time delay by $s = -\phi(\lambda)/\lambda$.

One can use the following argument for a general stationary process. The process (x_t) can be approximated by a sequence of harmonic processes:

$$x_t \approx \sum_{m=-M+1}^{M-1} e^{i\lambda_m^M t}(z(\lambda_{m+1}^M) - z(\lambda_m^M))$$

and accordingly, one obtains the following approximation of the output process:

$$y_t \approx \sum_{m=-M+1}^{M-1} e^{i\lambda_m^M t} k(\lambda_m^M)(z(\lambda_{m+1}^M) - z(\lambda_m^M))$$

$$= \sum_{m=-M+1}^{M-1} e^{i(\lambda_m^M t + \phi(\lambda_m^M))} r(\lambda_m^M)(z(\lambda_{m+1}^M) - z(\lambda_m^M))$$

which can be interpreted quite analogously to the above.

In case where the spectral densities exist, the interpretation is as follows. As discussed in Sect. 3.3, $f_x(\lambda)\Delta\lambda$ is a measure of the contribution of the oscillations in the frequency band $[\lambda, \lambda + \Delta\lambda]$ to the process and the same holds for $f_y(\lambda)\Delta\lambda$. According to the formula (4.10), it holds that

$$f_y(\lambda)\Delta\lambda = |k(\lambda)|^2 f_x(\lambda)\Delta\lambda.$$

These oscillations are hence

- attenuated for $|k(\lambda)| < 1$ and
- amplified for $|k(\lambda)| > 1$.

This means that certain oscillations can be amplified or attenuated with the help of the filter.

Lag operator L

The transfer function of the lag operator L is $k(\lambda) = e^{-i\lambda}$ and the amplitude response $r(\lambda) = |e^{-i\lambda}| = 1$ is constant equal to one. The lag operator hence does not change the "amplitudes" of the oscillations. The phase shift is $\phi(\lambda) = -\lambda$, corresponding to the time lag of one time unit ($s = -\phi(\lambda)/\lambda = 1$).

Difference filter $\Delta = (1 - L)$

The transfer function of the *difference filter* $\Delta = (1 - L)$ is $k(\lambda) = (1 - e^{-i\lambda}) = (1 - \cos(\lambda) + i\sin(\lambda))$ and the amplitude response is $|1 - e^{-i\lambda}| = \sqrt{2 - 2\cos(\lambda)}$. Slow oscillations ($\lambda \approx 0$) are thus attenuated, while oscillations with high frequencies $\lambda \approx \pi$ are amplified. This filter is therefore often used in time series analysis for trend adjustment. The phase response is $\phi(\lambda) = ((2\pi - \lambda) \mod 2\pi) - \pi)/2$.

Many economic time series, such as unemployment figures, show annual or seasonal patterns in addition to trend. For quarterly data, for example, the filter $(1 - L^4)$ is often used to eliminate both trend and seasonal components. Then $r(\lambda) = |1 - e^{-i4\lambda}| = \sqrt{2 - 2\cos(4\lambda)}$ and $\phi(\lambda) = ((2\pi - 4\lambda) \mod 2\pi) - \pi)/2$ are amplitude and phase response, respectively, for this filter. See Fig. 4.1.

Trend and seasonal adjustment

Such filters can be used, as already mentioned above, for trend or seasonal adjustment in particular. The transfer function is an important tool for analyzing the properties of these filters. Figure 4.2 illustrates two examples of filters for trend adjustment (of quarterly data). The amplitude response should be as small as possible for frequencies around zero and $r(\lambda) \approx 1$ should hold for all other frequencies. One wants $\phi(\lambda) = 0$

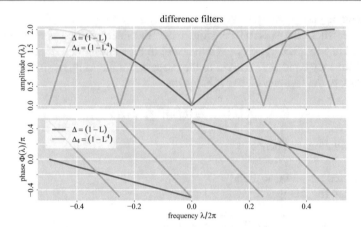

Fig. 4.1 Transfer function (amplitude and phase response) of the difference filters $\Delta = (1 - L)$ and $\Delta_4 = (1 - L^4)$

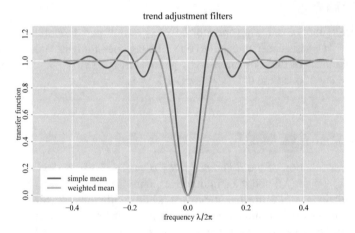

Fig. 4.2 Transfer function of two filters for trend adjustment. Simple mean: $\underline{a}(L) = L^0 - (0.5\,L^{-8} + L^{-7} + \cdots + L^7 + 0.5\,L^8)/16$; and "weighted" mean: $\underline{a}(L) = L^0 - \sum_{j=-8}^{8} w_j\,L^j$ with $w_j = (1 - |j/9|^3)^3 / \left(\sum_{j=-8}^{8} (1 - |j/9|^3)^3 \right)$. The simple mean is somewhat sharper around zero, but has larger maxima

for the phase response, since shifted cycles in the filtered version of the time series are not desirable. This last requirement can only be achieved with symmetric filters $\sum_j a_j\,L^j, a_j = a_{-j}$, that is, with filters that have a symmetric weight function.

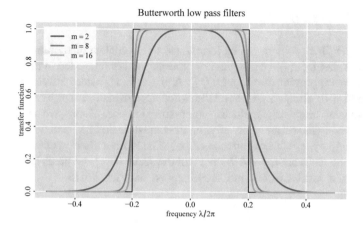

Fig. 4.3 Butterworth low-pass filters (of the order m) are characterized by the transfer function $b_m(\lambda) = |1 + e^{-i\lambda}|^{2m}/(|1 + e^{-i\lambda}|^{2m} + c|1 - e^{-i\lambda}|^{2m})$. These filters are rational (see Sect. 4.5) and have a symmetric weight function. The figure above shows how well Butterworth filters of the orders $m = 2, 8, 16$ approximate an ideal low-pass filter $(k(\lambda) = \mathbb{1}_{[-0.4\pi, 0.4\pi]}(\lambda))$. (The parameter c is chosen for each order m such that $b_m(0.4\pi) = 0.5$ holds.)

Low-pass, high-pass and band-pass filters

In signal processing, it is often necessary to have filters that delete certain frequency ranges as completely as possible, while allowing others to pass through as undisturbed as possible. The following idealized filters are of particular importance:

- Low-pass: $k(\lambda) = \mathbb{1}_{[-\lambda_0, \lambda_0]}(\lambda)$.
- High-pass: $k(\lambda) = 1 - \mathbb{1}_{[-\lambda_0, \lambda_0]}(\lambda)$.
- Band-pass: $k(\lambda) = \mathbb{1}_{[-\lambda_2, -\lambda_1]}(\lambda) + \mathbb{1}_{[\lambda_1, \lambda_2]}(\lambda)$.

The transfer functions of these idealized filters are not continuous; for this reason, they cannot be realized with an l_1-filter. However, these idealized transfer functions can be approximated with arbitrary accuracy using l_1 filters. See Fig. 4.3.

Exercise 4.12 Show the following assertion for scalar l_1 filters $\underline{k}(L) = \sum_j k_j L^j$: The transfer function $k(\lambda) = \sum_j k_j e^{-i\lambda j}$ is real-valued and symmetric $(k(\lambda) = k(-\lambda) \in \mathbb{R}, \forall \lambda \in [-\pi, \pi])$ if and only if the weight function is symmetric and real-valued $(k_j = k_{-j} \in \mathbb{R}, \forall j \in \mathbb{Z})$.

4.4   The Wiener Filter

The Wiener filter problem can be formulated in the context of stationary processes as follows: Let $(x'_t, y'_t)'$ be a stationary process. We are interested in the best linear least squares approximation of y_t by (x_t). A distinction is made between the approximation of y_t by all x_{t-j}, $j \in \mathbb{Z}$ and the causal approximation of y_t by x_{t-j}, $j \in \mathbb{N}_0$. Here, we want to investigate the first approximation only. Hence, we are concerned with projecting the components of y_t on the time domain $\mathbb{H}(x)$ and with describing this projection $\hat{y}_t$ as a linear function of (x_t). The Wiener filter was developed independently by Kolmogorov and Wiener (for the scalar case) and published in Wiener (1949). For a detailed discussion, see Lindquist and Picci (2015).

Suppose

$$\hat{y}_t = \mathrm{P}_{\mathbb{H}(x)}\, y_t = \underset{q \to \infty}{\mathrm{l.i.m}} \sum_{j=-q}^{q} l^q_j x_{t-j} \tag{4.18}$$

and

$$y_t = \hat{y}_t + u_t, \ u_t = (y_t - \hat{y}_t) \perp \mathbb{H}(x). \tag{4.19}$$

Here, (4.18) is a linear transformation of (x_t) and the approximation error u_t is orthogonal to $\mathbb{H}(x)$. Now let U be the forward shift associated with the process $(x'_t, y'_t)'$ on the common time domain $\mathbb{H}(x, y)$. Then it holds that

$$y_{t+1} = \mathrm{U}\, y_t = \mathrm{U}\, \hat{y}_t + \mathrm{U}\, u_t$$

where $\mathrm{U}\, \hat{y}_t = \mathrm{l.i.m}_q \sum_{j=-q}^{q} l^q_j x_{t+1-j} \in \mathbb{H}(x)$ and $\mathrm{U}\, u_t \perp \mathbb{H}(x)$. Hence, according to the projection theorem, the projection of y_{t+1} onto $\mathbb{H}(x)$ is $\hat{y}_{t+1} := \mathrm{U}\, \hat{y}_t$ and $u_{t+1} := \mathrm{U}\, u_t$ is the corresponding perpendicular. The projection $y_t \mapsto \hat{y}_t = \mathrm{P}_{\mathbb{H}(x)}\, y_t$ is thus a time-invariant transformation and therefore the weight function (l^q_j) does not depend on t. It also holds that

$$\gamma_{yx}(k) = \mathbf{E}(\hat{y}_{t+k} + u_{t+k})x'_t = \mathbf{E}\hat{y}_{t+k}x'_t = \gamma_{\hat{y}x}(k)$$

and therefore, in evident notation $f_{yx}(\lambda) = f_{\hat{y}x}(\lambda) = l(\lambda)f_x(\lambda)$, if the spectral densities exist. Thus, the following proposition follows from (4.9).

Proposition 4.13 (Wiener filter) *Let $(x'_t, y'_t)'$ be a stationary process with spectral density $\begin{pmatrix} f_x & f_{xy} \\ f_{yx} & f_y \end{pmatrix}$. If $f_x(\lambda) > 0$ μ-a.e., the transfer function $l(\lambda)$ of the best linear least squares approximation of y_t by (x_t) is determined by*

$$l(\lambda) = f_{yx}(\lambda)f_x^{-1}(\lambda). \tag{4.20}$$

The filter obtained from (4.20) is called *Wiener filter*. In analogy to the remark following Proposition 4.3, (4.20) can also be derived with f_x^τ and f_{yx}^τ. Thus, the existence of the spectral densities is not needed.

Moreover, the condition $f_x(\lambda) > 0$ is not necessary either. The essential condition for the Wiener filter is

$$l(\lambda) f_x(\lambda) = f_{yx}(\lambda).$$

This equation is μ-a.e. solvable because the matrix

$$\begin{pmatrix} f_x(\lambda) & f_{xy}(\lambda) \\ f_{yx}(\lambda) & f_y(\lambda) \end{pmatrix}$$

is μ-a.e. positive semidefinite. Every measurable solution $l(\lambda)$ is square integrable, i.e.

$$\int_{-\pi}^{\pi} l(\lambda) f_x(\lambda) l^*(\lambda) d\lambda < \infty$$

because $l(\lambda) f_x(\lambda) l^*(\lambda) \le f_y(\lambda)$ holds for all $\lambda \in [-\pi, \pi]$ and thus corresponds to a transfer function. However, the best approximation $\hat{y}_t$ is unique, regardless of the choice of $l(\lambda)$.

For instance, if

$$\int_{-\pi}^{\pi} \mathrm{tr}(l(\lambda) l(\lambda)^*) d\lambda < \infty$$

then the coefficients of the weight function are determined by

$$l_j = \frac{1}{2\pi} \int_{-\pi}^{\pi} l(\lambda) e^{i\lambda j} d\lambda.$$

If one interprets (x_t) as input and (y_t) as output, one obtains the following "noisy" system:

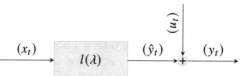

where the "disturbances" (u_t) are the part of the "noisy" output (y_t) which is not (linearly) explained by the inputs (x_t). In a way, the approach here is to solve an "inverse problem", i.e. one does not determine the output for given input (x_t) and given filter, but one determines the underlying (unknown) system from the second moments of the observed process $(x_t', y_t')'$, consisting of an input and (noisy) output. The condition $f_x(\lambda) > 0$ μ-a.e. is called a *persistent excitation condition*, as it implies that the inputs "vary sufficiently" in order to uniquely determine the transfer function of the system (via (4.20)).

By using the Wiener filter, we now can also give an interpretation of the cross-spectra. For this purpose, we consider a bivariate process $(x_t, y_t)'$ with a spectral density and assume that the Wiener filter is an l_1-filter. From the polar coordinate representation of the cross-spectrum

$$f_{yx}(\lambda) = |f_{yx}(\lambda)|e^{i\phi(\lambda)},$$

it follows that

$$l(\lambda) = \frac{f_{yx}(\lambda)}{f_x(\lambda)} = \frac{|f_{yx}(\lambda)|}{f_x(\lambda)}e^{i\phi(\lambda)}.$$

Thus, the amplitude response of the Wiener filter is equal to $r(\lambda) = \frac{|f_{yx}(\lambda)|}{f_x(\lambda)}$ and the phase response $\phi(\lambda)$ is equal to the "phase response" of the cross-spectrum f_{yx}. The least squares approximation $\hat{y}_t$ of y_t by $\{x_t, \ t \in \mathbb{Z}\}$ is obtained with the Wiener filter. Oscillations in an (infinitesimal) frequency band around (λ) are amplified or attenuated with the factor $r(\lambda)$ and phase-shifted by $\phi(\lambda) = \Im(\log(f_{yx}(\lambda)))$.

The decomposition

$$f_y(\lambda) = l(\lambda)f_x(\lambda)l(\lambda)^* + f_u(\lambda) = f_{\hat{y}}(\lambda) + f_u(\lambda) \tag{4.21}$$

where f_u, $f_{\hat{y}}$ denote the spectrum of the disturbances (u_t) and of the approximation $(\hat{y}_t)$ resp., is a frequency-specific measure of the quality of the approximation of the (noisy) output by (x_t) via the system. The *coherence* between (x_t) and (y_t) is defined by

$$C(\lambda) = \frac{f_{\hat{y}}(\lambda)}{f_y(\lambda)} = \frac{l(\lambda)f_x(\lambda)l(\lambda)^*}{f_y(\lambda)} = \frac{|f_{yx}(\lambda)|^2}{f_x(\lambda)f_y(\lambda)}.$$

It holds that $0 \leq C(\lambda) \leq 1$, as can be easily seen. The coherence can be interpreted as a frequency-specific coefficient of determination (squared correlation coefficient, respectively).

4.5   Rational Filters

So-called rational filters are of particular practical importance. They correspond to the stationary solutions of linear difference equations and can be described by finitely many parameters. A significant advantage of rational filters is that operations with these filters, such as series connection and inversion, can be traced back to algebraic operations.

A *rational matrix* is a matrix whose elements are rational functions (of a complex variable z). Due to the fact that rational functions form a field just like real and complex numbers, rational matrices share, from an abstract point of view, many properties with real or complex matrices. For a rational matrix $\underline{k}(z)$, we set

$$\underline{k}^*(z) = \left[\underline{k}(\bar{z}^{-1})\right]^*.$$

Thus, $\underline{k}^*$ is also rational and it holds that $\underline{k}^*(e^{-i\lambda}) = \left[\underline{k}(e^{-i\lambda})\right]^*$, i.e. $\underline{k}^*(e^{-i\lambda})$ is the Hermitian transpose of $\underline{k}(e^{-i\lambda})$. In most cases, we deal with rational matrices with real coefficients; in this case it holds that $\underline{k}^*(z) = \left[\underline{k}(z^{-1})\right]'$. If $\underline{k}$ is square and non-singular (as a rational matrix), we write $\underline{k}^{-*} = (\underline{k}^*)^{-1} = (\underline{k}^{-1})^*$.

Definition 4.14 Let $\underline{k}\colon \mathbb{C} \longrightarrow \mathbb{C}^{m \times n}$ be a rational matrix and $k\colon [-\pi, \pi] \longrightarrow \mathbb{C}^{m \times n}$. We say $\underline{k}$ is a *rational extension* of k if $k(\lambda) = \underline{k}(e^{-i\lambda})\ \forall \lambda \in [-\pi, \pi]$. In brief, we hence also say that k is *rational*.

It is quite easy to see that the rational extension is unique. If a transfer function has a rational extension, then we call the transfer function and the corresponding filter rational. In analogy, a spectral density with rational extension is also called rational. In the following, we will simply refer to the rational extension of a transfer function (or spectral density) as a transfer function (or spectral density) as well.

Example (rational filter)

The transfer function of a filter with finitely many coefficients $\underline{k}(L) = \sum_{j=-q_1}^{q_2} k_j L^j$, $q_i \in \mathbb{N}_0$ has a rational extension $\underline{k}(z) = \sum_{j=-q_1}^{q_2} k_j z^j$. The rational extension is a polynomial matrix if $q_1 = 0$, i.e. for filters of the form $\sum_{j=0}^{q} k_j L^j$.

Example (rational spectral density)

Let $\underline{k}(L) = \sum_j k_j L^j$ be an l_1-filter, whose transfer function has the rational extension $\underline{k}(z)$. In that case, the spectral density of the MA(∞) process $(x_t) = \underline{k}(L)(\epsilon_t)$ also has a rational extension: $\underline{f}(z) = \frac{1}{2\pi}\underline{k}(z)\Sigma\underline{k}^*(z)$.

Examples (operations with rational filters)

The product (and sum) of two rational transfer functions is again rational. Thus, the series connection (and the parallel connection) of two rational l_1-filters is also rational. Moreover, the inverse l_1-filter of a rational filter (if it exists) is rational, as we will show at the end of this section.

The Smith–McMillan form is a useful canonical representation of rational matrices (see, for example, Hannan and Deistler (2012)). For this, we first need the notion of a unimodular polynomial matrix: A square polynomial matrix $\underline{a}(z)$ is called *unimodular*, if $\det \underline{a}(z) \equiv \text{const} \neq 0$. The following example shows that this condition is equivalent to the assertion that $\underline{a}^{-1}(z)$ is a polynomial matrix, too.

Example (inverse of unimodular matrices)

In general, the inverse of a polynomial matrix $\underline{a}(z)$ is a rational matrix. Only for the case $\det \underline{a}(z) = d_0 \neq 0$ for all $z \in \mathbb{C}$, we obtain

$$\underline{a}^{-1}(z) = \frac{1}{d_0} \operatorname{adj} \underline{a}(z),$$

i.e. a polynomial inverse. Here, $\operatorname{adj} \underline{a}(z)$ denotes the adjugate matrix. The matrix

$$\underline{a}(z) = \begin{pmatrix} 1 & 0 \\ 0 & 1 \end{pmatrix} + \begin{pmatrix} \alpha & \alpha\beta \\ -\alpha\beta^{-1} & -\alpha \end{pmatrix} z = \begin{pmatrix} 1 + \alpha z & \alpha\beta z \\ -\alpha\beta^{-1} z & 1 - \alpha z \end{pmatrix}$$

e.g. is unimodular for all α, $\beta \in \mathbb{R}$, $\beta \neq 0$. The inverse is

$$\underline{a}^{-1}(z) = \begin{pmatrix} 1 - \alpha z & -\alpha\beta z \\ \alpha\beta^{-1} z & 1 + \alpha z \end{pmatrix}.$$

Proposition 4.15 (*Smith–McMillan form*) *Any rational $n \times n$ matrix $\underline{k}(z)$, whose determinant is not identically zero* ($\det \underline{k}(z) \not\equiv 0$), *has a representation of the form*

$$\underline{k}(z) = \underline{u}(z)\underline{\Lambda}(z)\underline{v}(z) \tag{4.22}$$

where $\underline{u}(z)$ and $\underline{v}(z)$ are unimodular polynomial matrices and $\underline{\Lambda}(z)$ is a diagonal matrix with diagonal elements $d_{ii} = p_{ii}/q_{ii}$. The polynomials p_{ii} and q_{ii} are relatively prime, p_{ii} and q_{ii} are monic (that is, the coefficients of the highest order term are normalized to 1), p_{ii} divides $p_{i+1,i+1}$ and $q_{i+1,i+1}$ divides q_{ii}. The diagonal matrix $\underline{\Lambda}$ is uniquely determined; however, the unimodular matrices $\underline{u}(z)$ and $\underline{v}(z)$ are, in general, not uniquely determined.

The Smith–McMillan form may be more generally defined for arbitrary rational $m \times n$-dimensional matrices. With the help of the Smith–McMillan form, we can now also define the poles and zeros of rational (square) matrices.

Definition 4.16 The zeros of $\underline{k}(z)$ are the zeros of the numerator polynomials p_{ii}, $i = 1, \ldots n$ and the poles of $\underline{k}(z)$ are the zeros of the denominator polynomials q_{ii}, $i = 1, \ldots, n$ in the Smith–McMillan form of $\underline{k}(z)$.

A complex number z_0 is a pole of $\underline{k}(z)$ if and only if z_0 is a pole of at least one element $k_{ij}(z)$. A complex number z_0 is the zero of $\underline{k}(z)$ if z_0 is a pole of the inverse $\underline{k}^{-1}(z)$. If z_0 is not a pole, z_0 is a zero of $\underline{k}(z)$ if and only if it holds that $\operatorname{rk}(\underline{k}(z_0)) < n$.

Exercise 4.17 Find an example of a rational matrix $\underline{k}(z)$, for which a complex number z_0 is both zero and pole.

Of course, the rational extension $\underline{k}(z)$ of the transfer function $k(\lambda)$ of an l_1-filter has no poles on the unit circle ($|z| = 1$). Conversely, we now show that any rational matrix $\underline{k}(z)$, which has no poles on the unit circle, can be interpreted as the rational extension of the transfer function of an l_1-filter. We first consider the scalar case

$$\underline{k}(z) = \frac{b(z)}{a(z)}$$

where $a(z) = a_0 + a_1 z + \cdots + a_{p-1} z^p + 1 z^p$ and $b(z) = b_0 + b_1 z + \cdots + b_q z^q$ are two relatively prime scalar polynomials. To show that $\underline{k}(z)$ is the transfer function of an l_1-filter, we first construct a Laurent series expansion of $\underline{k}(z)$ with absolutely summable coefficients. According to the fundamental theorem of algebra, $a(z)$ has a factorization

$$a(z) = \prod_{j=1}^{p} (z - z_j)$$

where z_j are the zeros of $a(z)$. We have assumed that $\underline{k}(z)$ has no pole on the unit circle, which means that $|z_j| \neq 1$ holds for $j = 1, \ldots, p$. The reciprocals of the factors $(z - z_j)$ can be developed using the geometric series:

$$\frac{1}{(z - z_j)} = \begin{cases} z^{-1} & \text{for } z_j = 0, \ |z| > 0 \\ z_j^{-1} \sum_{s=-1}^{-\infty} z_j^{-s} z^s & \text{for } 0 < |z_j| < 1, \ |z| > |z_j| \ . \\ -z_j^{-1} \sum_{s=0}^{\infty} z_j^{-s} z^s & \text{for } 1 < |z_j|, \ |z| < |z_j| \end{cases} \quad (4.23)$$

Thus, $\underline{k}(z)$ also has a Laurent series expansion

$$\underline{k}(z) = b(z) z^{-p_0} \prod_{j=p_0+1}^{p_0+p_1} \left[z_j^{-1} \sum_{s=-1}^{-\infty} z_j^{-s} z^s \right] \prod_{j=p_0+p_1+1}^{p} \left[-z_j^{-1} \sum_{s=0}^{\infty} z_j^{-s} z^s \right], \quad (4.24)$$

where the zeros are ordered by $z_j = 0$ for $1 \leq j \leq p_0$, $0 < |z_j| < 1$ for $p_0 + 1 \leq j \leq p_0 + p_1$ and $1 < |z_j|$ for $p_0 + p_1 < j \leq p$.

In the multivariate case, when $\underline{k}(z)$ is a rational matrix, we construct a series expansion for each element of the matrix according to this scheme, obtaining a Laurent series expansion of the matrix $\underline{k}(z) = \sum_{j=-\infty}^{\infty} k_j z^j$ that converges on an annulus of the form $\rho_1 < |z| < \rho_2$, where

$$\rho_1 = \max \left\{ |z| \mid z \text{ is a pole of } \underline{k} \text{ and } |z| < 1 \right\}$$
$$\rho_2 = \min \left\{ |z| \mid z \text{ is a pole of } \underline{k} \text{ and } |z| > 1 \right\}.$$

The coefficients k_j converge geometrically to zero, i.e. for each ρ with $\rho_1 < \rho < 1 < \rho^{-1} < \rho_2$ there exists a constant $c > 0$ such that

$$\|k_j\| \leq c \rho^{|j|} \ \forall j \in \mathbb{Z}. \quad (4.25)$$

In particular, the coefficients k_j are hence absolutely summable, and are therefore the weight function of an l_1-filter $\underline{k}(L) = \sum_j k_j L^j$. This construction shows that the filter is causal exactly when $\underline{k}(z)$ has no poles in or on the unit circle. In summarizing, we have the following.

Proposition 4.18 *A rational matrix $\underline{k}(z)$ is the transfer function of an l_1-filter if and only if $\underline{k}(z)$ has no poles on the unit circle. The filter coefficients are determined by the Laurent series expansion of $\underline{k}(z)$ (in particular, see (4.24)). The filter is causal if and only if $\underline{k}(z)$ has no poles in or on the unit circle.*

In conclusion, we briefly discuss the construction or computation of the inverse of a rational filter.

Proposition 4.19 *Let $\underline{k}(L)$ be a rational and square $(m = n)$ l_1-filter. The inverse filter to $\underline{k}(L)$ exists if and only if $\det \underline{k}(z) \neq 0 \, \forall |z| = 1$ holds for the transfer function. The inverse filter, if it exists, is l_1 and rational. The inverse filter is causal if and only if $\det \underline{k}(z) \neq 0 \, \forall |z| \leq 1$.*

Proof The condition $\det \underline{k}(z) \neq 0 \, \forall |z| = 1$ is necessary for the existence of the inverse l_1-filter, as shown in (4.16). If this condition is satisfied, then the inverse matrix $\underline{k}^{-1}(z)$ exists and is of the form

$$\underline{k}^{-1}(z) = \frac{1}{\det \underline{k}(z)} \operatorname{adj} \underline{k}(z).$$

The rational matrix $\underline{k}^{-1}(z)$ has no poles on the unit circle and is hence the (rational) transfer function of the desired inverse l_1-filter.   □

Example (difference filter is not invertible)

The difference filter $\Delta = (1 - L)$ is not invertible because the transfer function $(1 - z)$ is equal to zero for $z = 1$. One can also argue as follows: Let (x_t) be a stationary process with $\mathbf{E}x_t = \mu \neq 0$. The filtered process $y_t = \Delta x_t = x_t - x_{t-1}$ has the expectation $\mathbf{E}y_t = 0$. As a result of the formation of "differences", the information about the mean of x_t has been lost and there is no possibility to determine the expectation of x_t using the filtered process $(y_t = x_t - x_{t-1})$ only.

Exercise 4.20 Let (x_t) be a centered process with the autocovariance function $\gamma_x(k)$ and the spectral density $f_x(\lambda)$. Show for the difference process $(y_t) = (I - L)(x_t)$:

$$f_y(0) = 0 \quad \text{and} \quad \sum_{k=-\infty}^{\infty} \gamma_y(k) = 0.$$

Example (inverse of polynomial matrix)

A simple example for calculating the inverse of a polynomial matrix:

$$\underline{a}(z) = \begin{pmatrix} 1 & 0 \\ 0 & 1 \end{pmatrix} + \begin{pmatrix} a_{11} & a_{12} \\ a_{21} & a_{22} \end{pmatrix} z = \begin{pmatrix} 1 + a_{11}z & +a_{12}z \\ +a_{21}z & 1 + a_{22}z \end{pmatrix}.$$

The inverse of $\underline{a}(z)$ is

$$\begin{aligned}
\underline{a}^{-1}(z) &= \frac{1}{\det \underline{a}(z)} \operatorname{adj} \underline{a}(z) \\
&= \frac{1}{(1 + a_{11}z)(1 + a_{22}z) - (a_{12}z)(a_{21}z)} \begin{pmatrix} 1 + a_{22}z & -a_{12}z \\ -a_{21}z & 1 + a_{11}z \end{pmatrix} \\
&= \frac{1}{1 + (a_{11} + a_{22})z + (a_{11}a_{22} - a_{12}a_{21})z^2} \begin{pmatrix} 1 + a_{22}z & -a_{12}z \\ -a_{21}z & 1 + a_{11}z \end{pmatrix} \\
&= \begin{pmatrix} \frac{1+a_{22}z}{d(z)} & \frac{-a_{12}z}{d(z)} \\ \frac{-a_{21}z}{d(z)} & \frac{1+a_{11}z}{d(z)} \end{pmatrix}
\end{aligned}$$

where $d(z) = \det \underline{a}(z) = 1 + (a_{11} + a_{22})z + (a_{11}a_{22} - a_{12}a_{21})z^2$.

Definition 4.21 A square, rational and non-singular matrix $\underline{k}(z)$ is called *stable*, if $\underline{k}$ has no poles for $|z| \leq 1$. It is called *minimum-phase* (resp. *strictly minimum-phase*), if $\underline{k}$ has no zeros for $|z| < 1$ (resp. for $|z| \leq 1$).

If the matrix $\underline{k}$ is stable, then the (strict) minimum-phase condition is equivalent to $\det(\underline{k}(z)) \neq 0\ \forall |z| < 1$ resp. $\det(\underline{k}(z)) \neq 0\ \forall |z| \leq 1$. The corresponding filter is causal for a stable matrix $\underline{k}$. If the matrix is stable and strictly minimum-phase, then the inverse filter is causal and minimum-phase as well.

4.6   Difference Equations

Difference equations result, e.g. from the discretization of differential equations. Thus, the analysis of difference equations is of importance, e.g. in the technical sciences. For a detailed account, we refer to Deistler (1975). In principle, both observed and unobserved inputs can occur. This distinction does not matter in this section; later on, we will mainly discuss the case of unobserved inputs.

We consider systems of linear difference equations of the form

$$y_t = a_1 y_{t-1} + \cdots + a_p y_{t-p} + u_t \tag{4.26}$$

where $a_j \in \mathbb{R}^{n \times n}$ are parameter matrices and (u_t) is an n-dimensional input process. A *solution* on $\mathbb{Z}$ is a stochastic process (y_t) that, for given parameters a_j and given input (u_t), satisfies the equation (4.26) for all $t \in \mathbb{Z}$. It is easy to convince oneself that the following statement is correct:

Proposition 4.22 *The set of all solutions of (4.26) is of the following form: A (particular) solution of (4.26) plus the set of all solutions of the homogeneous equation*

$$y_t = a_1 y_{t-1} + \cdots + a_p y_{t-p}. \tag{4.27}$$

Using the lag operators L, we now write (4.26) (for $t \in \mathbb{Z}$) as

$$\underline{a}(L)(y_t) = (u_t)$$

where

$$\underline{a}(L) = I_n - a_1 L - \cdots - a_p L^p .$$

Unless specifically emphasized, we assume that (u_t) is stationary. If the filter $\underline{a}(L)$ is invertible, then

$$(y_t) = \underline{a}^{-1}(L)(u_t)$$

is a solution of (4.26). This solution is stationary and it is, as one can easily see, the only stationary solution.

Proposition 4.23 *If (u_t) is stationary and*

$$\det \underline{a}(z) \neq 0 \quad |z| = 1,$$

then the linear transformation

$$(y_t) = \underline{a}^{-1}(L)(u_t) = \left(\sum_{j=-\infty}^{\infty} k_j u_{t-j} \right) \tag{4.28}$$

yields a solution of (4.26). This solution is the only stationary solution. If it holds that

$$\det \underline{a}(z) \neq 0 \quad |z| \leq 1, \tag{4.29}$$

then this solution is causal. Thus $k_j = 0$ holds for $j < 0$.

Proof The transfer function $\underline{a}(z) = I_n - a_1 z - \cdots - a_p z^p$ of the filter $\underline{a}(L)$ is a polynomial matrix. Therefore, it follows immediately from Proposition 4.19 that the inverse filter $\underline{a}^{-1}(L)$ exists if and only if $\det \underline{a}(z) \neq 0$ $|z| = 1$ is satisfied. The inverse filter is causal if and only if $\det \underline{a}(z)$ has no zeros in (and on) the unit circle.

If the inverse filter exists, then $(y_t) := \underline{a}^{-1}(L)(u_t)$ is a (stationary) solution, since

$$\underline{a}(L)(y_t) = \underline{a}(L)\underline{a}^{-1}(L)(u_t) = (u_t).$$

Now if $(\tilde{y}_t)$ is an arbitrary stationary solution, the uniqueness of the stationary solution follows from

$$(\tilde{y}_t) = \underline{a}^{-1}(L)\underline{a}(L)(\tilde{y}_t) = \underline{a}^{-1}(L)(u_t) = \underline{a}^{-1}(L)\underline{a}(L)(y_t) = (y_t).$$

□

The coefficients of the l_1-filter $\underline{a}^{-1}(L)$ can be determined by comparison of coefficients from $\underline{a}(z)\underline{a}^{-1}(z) = I_n$ under the condition (4.29). One obtains the following recursive system of equations for the k_j's:

$$\begin{aligned} k_0 &= I_n \\ -a_1 k_0 + k_1 &= 0 \\ -a_2 k_0 - a_1 k_1 + k_2 &= 0 \\ \vdots \quad \vdots \end{aligned} \tag{4.30}$$

The condition (4.29) is often called *stability condition*, since it guarantees the existence of a stationary solution (y_t) for stationary inputs (u_t) under the a priori assumption of causality. This stationary solution is also called *steady-state* solution. If (u_t) is a regular process with innovations (ϵ_t), then the solution (4.28) is also regular with the same innovation process under the stability condition (4.29). This is a direct consequence of the fact that both $\underline{a}(L)$ and $\underline{a}^{-1}(L)$ are causal.

Example

The difference equation

$$y_t = y_{t-1} + u_t$$

cannot be solved with the method described here. However, all solutions can be easily determined recursively:

$$y_t = \begin{cases} y_0 + \sum_{j=1}^{t} u_j & \text{for } t > 0 \\ y_0 & \text{for } t = 0 \\ y_0 - \sum_{j=t+1}^{0} u_j & \text{for } t < 0. \end{cases}$$

A series of special cases of (4.26) leads to the following important model classes:

(1) If $(u_t = \epsilon_t)$ is white noise, then (4.26) is an AR system. Such AR systems and the corresponding stationary solutions, which are called AR processes, are discussed in detail in Chap. 5.
(2) If $(u_t) = \underline{b}(L)(\epsilon_t)$, $\underline{b}(L) = \sum_{j=0}^{q} b_j L^j$ is an MA(q) process, then (4.26) is called an ARMA System. These are discussed in more detail in Chap. 6.
(3) In the two models mentioned above, it is assumed that the input process (u_t) or resp. the underlying white noise is not observed. In ARX models, on the other hand, one assumes that (u_t) is of the form

$$u_t = \sum_{j=0}^{r} d_j x_{t-j} + \epsilon_t,$$

where (ϵ_t) is unobserved white noise and (x_t) is an *observed* input process with $\mathbf{E}\epsilon_t x_s = 0$ for all $t, s \in \mathbb{Z}$. Hence, an ARX system is a difference equation of the form

$$\underline{a}(L)(y_t) = \underline{d}(L)(x_t) + (\epsilon_t), \quad \underline{d}(L) = \sum_{j=0}^{r} d_j L^j.$$

(4) Analogously,

$$\underline{a}(L)(y_t) = \underline{d}(L)(x_t) + \underline{b}(L)(\epsilon_t)$$

is called an ARMAX system. Here, too, $\mathbf{E}\epsilon_t x_s = 0$ is required for all $t, s \in \mathbb{Z}$. For more details on ARX and ARMAX systems, see Sect. 8.1.

References

M. Deistler, z-transform and identification of linear econometric models with autocorrelated errors. Metrika **22**, 13–25 (1975)

E.J. Hannan, M. Deistler, *The Statistical Theory of Linear Systems*. Classics in Applied Mathematics (SIAM, Philadelphia, 2012). Originally published: (Wiley, New York, 1988)

A. Lindquist, G. Picci, *Linear Stochastic Systems; a Geometric Approach to Modeling, Estimation and Identification*, Series in Contemporary Mathematics, vol. 1. (Springer, Berlin, 2015). (ISBN 978-3-662-45749-8; 3-662-45749-0)

D.J. Newman, Shorter notes: a simple proof of wiener's 1/f theorem. Proc. Am. Math. Soc. **48**(1), 264–265 (1975). ISSN 00029939, 10886826, http://www.jstor.org/stable/2040730

N. Wiener, *Extrapolation, Interpolation, and Smoothing of Stationary Time Series* (Wiley, New York, 1949)

Autoregressive Processes

5

In this chapter, we discuss so-called autoregressive processes, i.e. stationary solutions of difference equations of the form

$$x_t = a_1 x_{t-1} + \cdots + a_p x_{t-p} + \epsilon_t, \ \forall t \in \mathbb{Z}$$

where $(\epsilon_t) \sim \mathrm{WN}(\Sigma)$ is white noise. AR models are probably the most widely used class of models for practical applications of time series analysis. Autoregressive models enable us to model processes with an "infinite" memory (i.e. with a covariance function γ, for which $\gamma(k) \neq 0$ holds for arbitrarily large k) and, in contrast to general MA(∞) processes, with a finite number of parameters. AR models are, for instance, well suited for describing processes with pronounced peaks in the spectral density. These are processes with dominating, "almost periodic" components, which can be found in many applications, for example in electrocardiogram (ECG) signals, among others. Moreover, any regular process can be approximated with arbitrary accuracy by an AR process if the order p is chosen large enough.

Another important advantage of autoregressive processes is the simplicity of their prediction. Under the stability condition, the one-step ahead prediction from the infinite past is simply $\hat{x}_{t,1} = a_1 x_t + \cdots + a_p x_{t+1-p}$. This means that the least squares prediction depends only on the last p past values and the corresponding coefficients are exactly the coefficients of the AR model. Therefore, the AR model is an explicit description of the intertemporal dependence structure. The model decomposes x_t into the part determined by the past and the innovation. Last but not the least, the AR model can also be estimated in a very simple way, e.g. using the so-called Yule–Walker equations. The model can be interpreted as a regression model. This explains the name "autoregressive" and shows that the model can also be estimated with the ordinary least squares method.

In the first section, we shall briefly discuss the stationary solution of the AR system under the stability condition. We already did essential preliminary work on

© Springer Nature Switzerland AG 2022
M. Deistler and W. Scherrer, *Time Series Models*, Lecture Notes in Statistics 224,
https://doi.org/10.1007/978-3-031-13213-1_5

this in the previous chapter. Next, we shall cover the prediction of AR processes from finite or infinite past and discuss the main characteristics of the spectral density of AR processes. The penultimate section focuses on the Yule–Walker equations, which establish the interrelation between the parameters of the AR system and the covariance function. As stated above, these equations also form the basis for one of the most important methods used to estimate AR systems. Due to the vast number of applications of AR systems, an enormous number of algorithms was developed for their estimation. Special attention was paid to the development of recursive, numerically very efficient methods, which are very important, e.g. in real-time signal processing.

In the last section, we will omit the stability condition and briefly discuss the stationary solutions of AR systems in general. This section will also consider special non-stationary solutions, so-called integrated and cointegrated processes, which occur in the case of a so-called unit root. One of the most important results in this section is Granger's representation theorem.

Earlier references to AR processes can be found in Yule (1927)[1] and Mann and Wald (1943).[2] A detailed discussion (also covering the multivariate case) can be found in Anderson (1971),[3] Hannan (1970) and Lütkepohl (2005). A fundamental, earlier method of how to estimate parameters is the Durbin–Levinson algorithm; see Levinson (1947), Durbin (1960).[4, 5] Standard literature discussing the integrated case includes Engle and Granger (1987),[6] Johansen (1995) and Phillips (1987).

5.1   The Stability Condition

An *autoregressive system* (AR system) is a difference equation of the form

$$x_t = a_1 x_{t-1} + \cdots + a_p x_{t-p} + \epsilon_t, \ \forall t \in \mathbb{Z} \tag{5.1}$$

where $a_j \in \mathbb{R}^{n \times n}$, $a_p \neq 0$ and $(\epsilon_t) \sim \text{WN}(\Sigma)$. A stationary solution of (5.1), i.e. a stationary process (x_t), which satisfies the equation(s) for all $t \in \mathbb{Z}$, is a so-called *autoregressive process* of the order p (AR(p) process).

---

[1] George Udny Yule (1871–1951), Scottish statistician and one of the early pioneers of time series analysis. AR and MA models are based on his research.

[2] Abraham Wald (1902–1950), German-speaking U.S. mathematician, econometrician and statistician (born in Transylvania). Founder of statistical decision theory; numerous fundamental works including the Wald test and sequential testing procedures.

[3] Theodore W. Anderson (1918–2016), US-American statistician. One of the founders of modern time series analysis together with E. J. Hannan.

[4] Norman Levinson (1912–1975), US-American mathematician, particularly known in our context for the Durbin–Levinson algorithm.

[5] James Durbin (1923–2012), British statistician and econometrician, mainly known for the Durbin–Watson test and tests on structural breaks.

[6] Clive W. J. Granger (1934–2009), British-US econometrician. He worked on the spectral analysis of economic time series, on the analysis of causality ("Granger causality") and on cointegration. He was awarded the Nobel Prize in Economic Sciences in 2003.

The AR system is equivalent to the following "filter equation"

$$\underline{a}(L)(x_t) = (I_n - a_1 L - \cdots - a_p L^p)(x_t) = (\epsilon_t). \tag{5.2}$$

The transfer function

$$\underline{a}(z) = I_n - a_1 z - \cdots - a_p z^p \tag{5.3}$$

of the filter $\underline{a}(L) = I_n - a_1 L - \cdots - a_p L^p$ is a polynomial matrix of degree p and is often called the AR polynomial. We always impose the so-called *stability condition* in this chapter, unless the contrary is mentioned explicitly:

$$\det(\underline{a}(z)) \neq 0 \ \forall |z| \leq 1. \tag{5.4}$$

An AR system that satisfies this condition is called stable. As shown in Proposition 4.23, the AR system then has a unique stationary solution, and this solution is a causal MA(∞) process

$$(x_t) = \underline{a}^{-1}(L)(\epsilon_t) = \left(\sum_{j \geq 0} k_j L^j\right)(\epsilon_t) = \left(\sum_{j \geq 0} k_j \epsilon_{t-j}\right). \tag{5.5}$$

From this representation, it follows that ϵ_t is orthogonal to x_s, $s < t$. Hence, the AR system (5.1) is a regression model, which describes x_t by its *own* past $(x_{t-1}, \ldots, x_{t-p})$ and an error orthogonal to it. This observation explains the name "autoregressive process". As we shall see in the following sections, orthogonality is also key to the prediction of AR processes and to the so-called Yule–Walker equations, and thus to the estimation of AR systems.

The coefficients of the inverse filter $\underline{a}^{-1}(L)$ can be computed recursively, as described in (4.30). In particular, it holds that

$$k_0 = I_n. \tag{5.6}$$

The coefficients converge to zero at a geometric rate, i.e. $\|k_j\| \leq c\rho^j$ for a $c < \infty$ and $0 < \rho < 1$. For any (square integrable) solution $(\tilde{x}_t)$ of the AR system, it also holds that

$$\underset{t \to \infty}{\text{l.i.m}}(x_t - \tilde{x}_t) = 0$$

and this convergence is so fast that it does not matter for the asymptotics of typical estimators whether one considers an arbitrary solution $(\tilde{x}_t)$ or the stationary solution (x_t).

For the analysis of AR systems or AR processes, it is often favorable to consider the "stacked" process $x_t^p = (x_t', \ldots, x_{t+1-p}')'$. One can easily see that (x_t) is a solution of (5.1) if and only if the stacked process is a solution of the AR(1) system

$$x_t^p = A x_{t-1}^p + B\epsilon_t, \tag{5.7}$$

where

$$B = \begin{pmatrix} I_n \\ 0 \\ 0 \\ \vdots \\ 0 \end{pmatrix} \in \mathbb{R}^{pn \times n}, \quad A = \begin{pmatrix} a_1 & a_2 & \cdots & a_{p-1} & a_p \\ I_n & 0 & \cdots & 0 & 0 \\ 0 & I_n & \cdots & 0 & 0 \\ \vdots & \vdots & \ddots & \vdots & \vdots \\ 0 & 0 & \cdots & I_n & 0 \end{pmatrix} \in \mathbb{R}^{np \times np}. \qquad (5.8)$$

The matrix A is called the *companion matrix* of the polynomial matrix $\underline{a}(z)$. The eigenvalues of the companion matrix A are related to the zeros of $\det(\underline{a}(z))$ as follows.

Lemma 5.1 *For $z \in \mathbb{C}$, the following three statements are equivalent:*

(1) $\det(I - a_1 z - \cdots - a_p z^p) = 0$;
(2) $\det(I - Az) = 0$;
(3) $(1/z)$ is a nonzero eigenvalue of A.

Proof   (2) $\Longleftrightarrow$ (3): From $\det(I - Az) = 0$, it follows that $z \neq 0$. Hence, $\det(I - Az) = 0$ is equivalent to $\det(\frac{1}{z}I - A) = 0$.

(1) $\Longleftrightarrow$ (3): Let $c = (c_1, \ldots, c_p) \in \mathbb{C}^{1 \times np}$, $\underline{c}(z) = c_1 + c_2 z + \cdots + c_p z^{p-1}$ and $\lambda \in \mathbb{C}$. The following equations are obtained by equivalence transformations:

$$cA = \lambda c$$
$$c_1(a_1, \ldots, a_p) + (c_2, \ldots, c_p, 0) = \lambda(c_1, \ldots, c_p)$$
$$c_1(I, -a_1, \ldots, -a_p) = (c_1, c_2, \ldots, c_p, 0) - \lambda(0, c_1, \ldots, c_p)$$
$$c_1 \underline{a}(z) = (1 - \lambda z)\underline{c}(z) \quad \forall z \in \mathbb{C}.$$

If $\lambda \neq 0$ is an eigenvalue of A and c an associated left eigenvector, then it follows that $c_1 \neq 0$ and $c_1 \underline{a}(1/\lambda) = 0$. That is, $z = 1/\lambda$ is a zero of $\det(\underline{a}(z))$.
Conversely, if $\det(\underline{a}(1/\lambda)) = 0$ then there exists a $c_1 \in \mathbb{C}^{1 \times n}$, $c_1 \neq 0$ such that $c_1 \underline{a}(1/\lambda) = 0$. Hence, there exists a polynomial $\underline{c}(z) = c_1 + \cdots + c_p z^{p-1}$, such that $c_1 \underline{a}(z) = (1 - \lambda z)\underline{c}(z)$ and $c = (c_1, \ldots, c_p)$ is thus a left eigenvector of A to the eigenvalue λ. $\square$

In particular, this lemma implies that the stability condition (5.4) for the AR(p) system is equivalent to the stability condition of the AR(1) system (5.7) for the stacked process (x_t^p). Thus, in principle, one can now trace the AR(p) case back to the AR(1) case.

Under the stability condition, it holds that $\varrho(A) < 1$, where $\varrho(A) = \max_i |\lambda_i(A)|$ denotes the *spectral radius* of A. Therefore, for any $\varrho(A) < \rho < 1$, there exists a constant $c \in \mathbb{R}_+$ such that $\|A^k\| \le c\rho^k$ holds for all $k \ge 0$.

Exercise 5.2 Show the following: The scalar AR(2) polynomial $a(z) = 1 - a_1 z - a_2 z^2$ satisfies the stability condition if and only if the coefficients (a_1, a_2) are contained in the triangle determined by the inequalities

$$|a_2| < 1$$
$$a_2 + a_1 < 1$$
$$a_2 - a_1 < 1.$$

Exercise 5.3 Let $(x_t \mid t \in \mathbb{Z})$ be a process and $(x_t^P \mid t \in \mathbb{Z})$ the associated stacked process. Show that (x_t) is stationary (and regular) if and only if (x_t^P) is stationary (and regular). If (ϵ_t) is the innovation process of (x_t), then $((\epsilon_t', 0, \ldots, 0)')$ is the innovation process of (x_t^P).

Exercise 5.4 Let (x_t) be the stationary solution of the AR system (5.1) and $(\tilde{x}_t)$ an arbitrary (square integrable) solution. Show that

$$\mathbf{E}((x_t - \tilde{x}_t)'(x_t - \tilde{x}_t)) \le c\rho^{2t}$$

for a suitable constant $c \in \mathbb{R}_+$. Note that $(x_t - \tilde{x}_t)$ is a solution of the homogenous equation; hence, the following holds for the stacked vectors:

$$(x_t^P - \tilde{x}_t^P) = A(x_{t-1}^P - \tilde{x}_{t-1}^P) = A^t(x_0^P - \tilde{x}_0^P).$$

Exercise 5.5 Show $\|k_j\| \le c\rho^j$ for a suitable constant $c \in \mathbb{R}_+$. (See also (4.25).) Note that from the recursion (4.30), it follows that

$$\begin{pmatrix} k_j \\ \vdots \\ k_{j+1-p} \end{pmatrix} = A \begin{pmatrix} k_{j-1} \\ \vdots \\ k_{j-p} \end{pmatrix} = A^{j-p+1} \begin{pmatrix} k_{p-1} \\ \vdots \\ k_0 \end{pmatrix}, \quad \text{for } j \ge p - 1.$$

Exercise 5.6 The components of an AR process are generally not AR processes. Construct an appropriate example, e.g. a bivariate AR(1) process $(x_t = (x_{1t}, x_{2t})'$, such that (x_{1t}) is not an AR process. Therefore, the set of AR processes is not closed with regard to marginalization.

Exercise 5.7 (*Harmonic processes and AR systems*) Show that (scalar) harmonic processes are also AR processes: Construct the associated AR system and show that $\Sigma = 0$ holds, i.e. the process is singular. Note: Using the notation of the subsection on harmonic processes in Sect. 1.4, $x_{t+1}^K = \Theta D^{t+1} z = \Theta D \Theta^{-1} \Theta D^t z = \Theta D \Theta^{-1} x_t^K$ holds, where $D = \text{diag}(\theta_{1-M}, \ldots, \theta_M)$. Also, show that this AR system does *not* satisfy the stability condition (except for the trivial process ($x_t = 0$)). In this case, the zeros of $\underline{a}(z)$ all lie on the unit circle.

As already mentioned above, we shall always assume the stability condition in the following three Sects. 5.2, 5.3 and 5.4.

5.2   Prediction

The prediction of AR processes is particularly simple. The AR representation (5.1) implies $\mathbb{H}_t(\epsilon) \subset \mathbb{H}_t(x)$. Due to the stability condition (5.4), $x_t = \sum_{j \geq 0} k_j \epsilon_{t-j}$ and thus $\mathbb{H}_t(x) \subset \mathbb{H}_t(\epsilon)$. Moreover, $k_0 = I_n$; see (5.6). By Corollary 2.14, (x_t) is hence regular and the (ϵ_t) are innovations of (x_t). Therefore, the following holds for the one-step ahead prediction (from the infinite past):

$$u_{t,1} = \epsilon_{t+1}$$
$$\Sigma_1 = \mathbb{E} u_{t,1} u'_{t,1} = \Sigma$$
$$\hat{x}_{t,1} = x_{t+1} - \epsilon_{t+1} = a_1 x_t + \cdots + a_p x_{t+1-p}.$$

The coefficients of the optimal one-step ahead prediction are thus exactly the coefficients of the (stable) AR representation and only the last p values are needed for prediction. This property is a *characteristic* of AR processes. This means that a stationary process (x_t) is an AR process if and only if the one-step ahead prediction from the infinite past depends only on *finitely* many values. Thus, for the prediction from the finite past, it follows that $\hat{x}_{t,1,k} = \hat{x}_{t,1}$ and $\Sigma_{1,k} = \Sigma_1$ as long as $k \geq p$ values are used for prediction.

Multistep ahead prediction can be determined recursively in a very simple way. Let $P = P_{\mathbb{H}_t(x)}$ be the projection onto the space $\mathbb{H}_t(x)$. For $h = 2$ and $h = 3$, we obtain, e.g.

$$\hat{x}_{t,2} = P(x_{t+2}) = a_1 \underbrace{P x_{t+1}}_{\hat{x}_{t,1}} + a_2 \underbrace{P x_t}_{x_t} \cdots + a_p \underbrace{P x_{t+2-p}}_{x_{t+2-p}} + \underbrace{P \epsilon_{t+2}}_{=0}$$
$$= a_1 \hat{x}_{t,1} + a_2 x_t + \cdots + a_p x_{t+2-p}$$
$$u_{t,2} = (x_{t+2} - \hat{x}_{t,2}) = \epsilon_{t+2} + a_1 u_{t,1} = \epsilon_{t+2} + a_1 \epsilon_{t+1}$$

$$\hat{x}_{t,3} = P(x_{t+3}) = a_1 \underbrace{P\,x_{t+2}}_{\hat{x}_{t,2}} + a_2 \underbrace{P\,x_{t+1}}_{\hat{x}_{t,1}} + a_3 \underbrace{P\,x_t}_{x_t} + \cdots + a_p \underbrace{P\,x_{t+3-p}}_{x_{t+3-p}} + \underbrace{P\,\epsilon_{t+3}}_{=0}$$

$$= a_1\hat{x}_{t,2} + a_2\hat{x}_{t,1} + a_3 x_t + \cdots + a_p x_{t+3-p}$$

$$u_{t,3} = (x_{t+3} - \hat{x}_{t,3}) = \epsilon_{t+3} + a_1 u_{t,2} + a_2 u_{t,1} = \epsilon_{t+3} + a_1\epsilon_{t+2} + (a_1^2 + a_2)\epsilon_{t+1}.$$

Thus, even for multistep ahead prediction, it suffices to use the last p values. It is also easy to convince oneself that the representation of the h-step ahead prediction errors obtained from this recursive procedure naturally coincide with the representation $u_{t,h} = \sum_{j=0}^{h-1} k_j \epsilon_{t+h-j}$ given in Corollary 2.14.

5.3 Spectral Density

Given the stability condition (5.4), the AR(p) process is regular and its spectral density exists and by Proposition 4.3 is of the form:

$$f(\lambda) = \frac{1}{2\pi} \underline{a}^{-1}(e^{-i\lambda}) \Sigma (\underline{a}^{-1}(e^{-i\lambda}))^*. \tag{5.9}$$

In the scalar case (with $\epsilon_t \sim WN(\sigma^2)$), one obtains a somewhat simpler representation

$$f(\lambda) = \frac{\sigma^2}{2\pi |\underline{a}(e^{-i\lambda})|^2}. \tag{5.10}$$

By (5.9) the spectral density f is rational; more precisely, f has the rational extension

$$\underline{f}(z) = \frac{1}{2\pi} \underline{a}^{-1}(z) \Sigma \underline{a}^{-*}(z).$$

In the scalar case ($n = 1$), $\underline{f}$ has the representation

$$\underline{f}(z) = \frac{\sigma^2}{2\pi} \frac{1}{\underline{a}(z)\underline{a}(\frac{1}{z})} = \frac{\sigma^2}{2\pi} \frac{z^p}{\underline{a}(z)\tilde{\underline{a}}(z)}$$

where $\sigma^2 = \Sigma$ and $\tilde{\underline{a}}(z) = z^p \underline{a}(z^{-1}) = (z^p - a_1 z^{p-1} - \cdots - a_p)$. For $a_p \neq 0$, one can immediately see that $\underline{a}(z) = 0$ is equivalent to $\tilde{\underline{a}}(z^{-1}) = 0$. The poles of f are the zeros of $\underline{a}(z)\tilde{\underline{a}}(z)$ and therefore, each zero z_k of $\underline{a}(z)$ "generates" two poles of $\underline{f}(z)$, namely z_k and the zero z_k^{-1} "mirrored" on the unit circle. Zeros of $\underline{a}(z)$, which are close to the unit circle ($|z| = 1$), lead to peaks in the spectrum (as a function of $\lambda \in [-\pi, \pi]$). However, if all zeros of $\underline{a}(z)$ are far away from the unit circle, then one obtains a relatively flat spectrum.

In the case of an AR(1) process $x_t = a_1 x_{t-1} + \epsilon_t$, the poles of $\underline{f}(z)$ equal $z_1 = 1/a_1$ and $z_1^{-1} = a_1$. That is, for a_1 close to one, one obtains a spectrum with a peak around the frequency $\lambda = 0$, i.e. (x_t) is dominated mainly by low-frequency oscillations, while the high frequencies (around $\lambda = \pi$) dominate for a_1 close to -1.

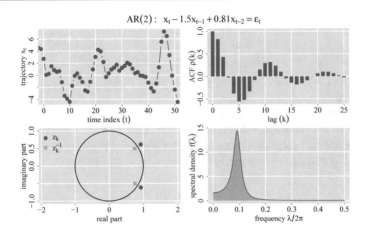

Fig. 5.1 AR(2) process $x_t = 1.5x_{t-1} - 0.81x_{t-2} + \epsilon_t$: The upper left panel shows a trajectory of this process, and the upper right panel shows the autocorrelation function. The bottom left panel shows the poles of $\underline{f}(z)$; the bottom right panel displays the spectral density

For an AR(2) process, either two real zeros or a pair of complex zeros $z_1 = \rho e^{i\lambda}$, $1 < \rho$, $\lambda \in (0, \pi)$ and $z_2 = \overline{z_1} = \rho e^{-i\lambda}$ can occur. In the latter case, $\underline{f}(z)$ has the poles $\rho e^{i\lambda}$, $\rho e^{-i\lambda}$, $\rho^{-1} e^{-i\lambda}$, $\rho^{-1} e^{i\lambda}$. If these poles are close to the unit circle (i.e. ρ close to one), then the spectrum has a peak at frequency λ (and $-\lambda$). Thus, the process is dominated primarily by oscillations with frequencies around λ. See Fig. 5.1.

From this discussion, one can see that an AR(p) model can be used very easily in order to construct spectra with pronounced peaks. AR(p) models are therefore particularly well-suited for modeling time series that have "almost" periodic components, e.g. audio signals. See also Fig. 5.2.

In the multivariate case, the poles of $\underline{f}$ are determined by the zeros of $\det \underline{a}(z)$: If $z_1, \ldots, z_r$ are the zeros of $\det \underline{a}(z)$, then $\underline{f}(z)$ has the poles $z_1, z_1^{-1}, \ldots, z_r, z_r^{-1}$. Here, too, the poles thus occur in real pairs $(z, z^{-1}) \in \mathbb{R}^2$ or resp. in quadruples $(z, \overline{z}, z^{-1}, (\overline{z})^{-1}) \in (\mathbb{C} \setminus \mathbb{R})^4$.

Exercise 5.8 Consider the following AR(4) system:

$$x_t = ax_{t-4} + \epsilon_t$$

where $|a| < 1$ and $(\epsilon_t) \sim \text{WN}(\sigma^2)$. Show that $\left(x_t = \sum_{j \geq 0} a^j \epsilon_{t-4j}\right)$ is the only stationary solution. Determine the autocorrelation function ρ and the spectral density f of (x_t). Sketch the ACF and the spectral density for the case $a = 0.9$. Also calculate the poles of the rational extension $\underline{f}(z)$ of the spectral density for this case.

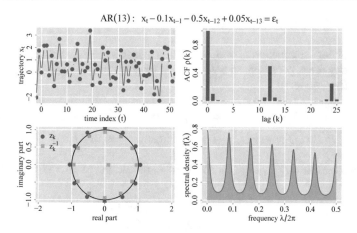

Fig. 5.2 AR(13) process $x_t = 0.1x_t + 0.5x_{t-12} - 0.05x_{t-13} + \epsilon_t$. The upper left panel shows a trajectory of this process, and the upper right panel shows the autocorrelation function. The bottom left panel shows the poles of $\underline{f}(z)$; the bottom right panel displays the spectral density

Exercise 5.9 Consider the following bivariate AR(1) system with $(\epsilon_t) \sim \text{WN}(\Sigma)$, $\Sigma = 2\pi I_2$:

$$x_t = \begin{pmatrix} 0.5 & 0.1 \\ -0.75 & 0.1 \end{pmatrix} x_{t-1} + \epsilon_t.$$

(1) Verify the stability condition.
(2) Calculate the covariance function $\gamma(k)$ of (x_t) for $k = 0, 1, 2$.
(3) Calculate $\underline{a}(z)^{-1}$ and the spectral density $\underline{f}(z)$ of (x_t).
(4) Plot the two autospectra $f_{11}(\lambda)$, $f_{22}(\lambda)$ and the coherence

$$C(\lambda) = \frac{|f_{21}(\lambda)|^2}{f_{11}(\lambda) f_{22}(\lambda)}.$$

(5) Now consider the linear least squares approximation of x_{2t} by x_{1t}:

$$\hat{x}_{2t} = (\gamma_{21}(0)\gamma_{11}(0)^{-1})x_{1t}.$$

Plot the spectral density of the approximation error ($\hat{u}_{2t} = (x_{2t} - \gamma_{21}(0)\gamma_{11}(0)^{-1}x_{1t})$).
(6) The Wiener filter yields the linear least squares approximation of x_{2t} by (x_{1t}). In contrast to the "static" approximation above, future and past values of (x_{1t}) are also used for the approximation. Plot the spectral density of the approximation error of the Wiener filter (and compare it with the above result). Note that the spectral density of the error follows from Eq. (4.21).

5.4   The Yule–Walker Equations

By multiplying the equation $x_t = a_1 x_{t-1} + \cdots + a_p x_{t-p} + \epsilon_t$ from the right by x'_{t-j} and forming the expectation on both sides of the equation, we obtain the so-called *Yule–Walker equations*:

$$\gamma(0) = a_1 \gamma(-1) + \cdots + a_p \gamma(-p) + \Sigma \quad \text{for } j = 0 \qquad (5.11)$$

$$\gamma(j) = a_1 \gamma(j-1) + \cdots + a_p \gamma(j-p) \quad \text{for } j > 0. \qquad (5.12)$$

Here, one uses $x_t = \sum_{j \geq 0} k_j \epsilon_{t-j}$ and $k_0 = I_n$ (according to Eq. (5.6)) and thus

$$\mathbf{E}\epsilon_t x'_{t-j} = \begin{cases} 0 & \text{for } j > 0, \\ \Sigma k'_0 = \Sigma & \text{for } j = 0, \\ \Sigma k'_j & \text{for } j < 0. \end{cases}$$

The Yule–Walker equations represent the relationship between the covariance function of the process and the parameters $(a_1, \ldots, a_p, \Sigma)$ of the underlying AR system. As we will show in the following, it is therefore possible to determine the covariance function for given parameters on the one hand and the parameters for given covariance functions on the other hand. By inserting the estimated autocovariances $\hat{\gamma}(k)$ into the Yule–Walker equations and solving the equations for the parameters, one obtains the so-called Yule–Walker estimators for the AR parameters.

The Covariance Function

First, we want to determine the covariance function $\gamma(\cdot)$ of the AR process (x_t). For this, we consider the stacked process (x_t^p) generated by the AR(1) system (5.7). The Yule–Walker equations for the stacked process (x_t^p) are

$$\Gamma_p = \Gamma_p(0) = A\Gamma_p(-1) + B\Sigma B' \qquad (5.13)$$

$$\Gamma_p(j) = A\Gamma_p(j-1) = A^j \Gamma_p \quad \text{for } j > 0 \qquad (5.14)$$

where $\Gamma_p(\cdot)$ denotes the covariance function of (x_t^p), i.e.

$$\Gamma_p(j) = \mathbf{E}x_{t+j}^p (x_t^p)' = (\gamma(j+l-k))_{k,l=1,\ldots,p}.$$

By inserting $\Gamma_p(-1) = \Gamma_p(1)' = \Gamma_p A'$ in (5.13), one obtains

$$\Gamma_p = A\Gamma_p A' + B\Sigma B' \qquad (5.15)$$

$$\Gamma_p(j) = A^j \Gamma_p \quad \text{for } j > 0. \qquad (5.16)$$

Equation (5.15) also follows from, e.g.

$$\Gamma_p = \mathbf{E}x_t^p (x_t^p)' = \mathbf{E}(Ax_{t-1}^p + B\epsilon_t)(Ax_{t-1}^p + B\epsilon_t)'$$
$$= A\mathbf{E}x_{t-1}^p (x_{t-1}^p)'A' + B\mathbf{E}\epsilon_t\epsilon_t'B' = A\Gamma_p A' + B\Sigma B'.$$

It is easy to convince oneself that Eqs. (5.15), (5.16) are (algebraically) equivalent to the Yule–Walker Eqs. (5.11) and (5.12). Since the spectral radius of A is less than one, (5.15) has a *unique* solution

$$\Gamma_p = \sum_{k\geq 0} A^k B\Sigma B'(A^k)'. \tag{5.17}$$

The autocovariances $\gamma(j)$ for $j \geq p$ can then be determined recursively from (5.16) and (5.12), respectively.

Exercise 5.10 Show that the covariance function of a regular AR process converges to zero at a geometric rate, i.e. $\|\gamma(k)\| \leq c\rho^k$ for suitable constants $c, \rho \in \mathbb{R}_+$, $\rho < 1$. Note: Since the AR process is regular, you can assume without loss of generality that the stability condition is satisfied. In this case, the result is obtained, e.g. with the help of the Yule–Walker equations, particularly (5.16).

The Yule–Walker Estimator

We now consider the Yule–Walker Eqs. (5.11), (5.12) as a system of equations for the AR parameter $a_1, \ldots, a_p$ and Σ. Typically, one just uses the equations for $j = 0$ and $j = 1, \ldots, p$. We thus obtain the following system of equations:

$$\gamma(0) = (a_1, \ldots, a_p)(\gamma(1), \ldots, \gamma(p))' + \Sigma \tag{5.18}$$
$$(\gamma(1), \ldots, \gamma(p)) = (a_1, \ldots, a_p)\Gamma_p. \tag{5.19}$$

In this section, $\gamma(\cdot)$ is the covariance function of an arbitrary stationary process, which does not necessarily have to be an AR process. In particular, the following cases are important for applications:

(1) The covariance function γ corresponds to an AR(p) process. Here, the question is whether one can (uniquely) determine the AR parameters from the covariance function γ.
(2) The covariance function corresponds to an AR(p_0) process, but $p_0 \neq p$.
(3) The process is stationary, but not necessarily an AR(p) process. One attempts to approximate the process (x_t) by an AR(p) process (or an AR(p) system, respectively).
(4) The covariance function is an empirical covariance function, which means that we insert estimates $\hat{\gamma}(k)$ into the Yule–Walker equations and then solve the equations for the parameters to obtain estimators for the AR parameters. These estimators are called *Yule–Walker estimators*.

In the following, $a_1, \ldots, a_p$, Σ denote solutions of the Yule–Walker Eqs. (5.18) and (5.19). These equations are exactly the "prediction equations" for the one-step ahead prediction from $k = p$ past values; see (2.4) and (2.5). We can therefore immediately conclude that the Yule–Walker Eqs. (5.18) and (5.19) are *always* solvable, provided that the sequence $\gamma(k)$ is a covariance function, i.e. positive semidefinite. The variance $\Sigma = \Sigma_{1,p}$ is uniquely determined and the AR coefficients $a_1, \ldots, a_p$ are uniquely determined from the Yule–Walker equations if and only if Γ_p is positive definite.

This does not necessarily mean that every stationary process is an AR process. To show that the covariance function γ corresponds to an AR(p) process with parameters $a_1, \ldots, a_p$, Σ, one still has to show that Eqs. (5.12) are also satisfied for all $j > p$.

Proposition 5.11 *If $(\gamma(k) \in \mathbb{R}^{n \times n} \mid k \in \mathbb{Z})$ is a positive semidefinite sequence, then the Yule–Walker equations are always solvable. The variance Σ is uniquely determined and the AR coefficients $a_1, \ldots, a_p$ are uniquely determined if and only if Γ_p is positive definite.*

For the case $\Gamma_{p+1} > 0$, the Yule–Walker equations yield AR parameters that satisfy the stability condition, i.e. $\det(I - a_1 z - \cdots - a_p z^p) \neq 0$ for all $|z| \leq 1$. In any case, a solution always exists for which $\det(I - a_1 z - \cdots - a_p z^p) \neq 0$ holds for all $|z| < 1$. One can therefore at least exclude zeros within the unit circle.

Proof We just need to prove the second part of the theorem. According to Lemma 5.1, we have to show that the spectral radius of the companion matrix A is always less than or equal to one and in the case of $\Gamma_{p+1} > 0$ less than one. Thus, let $c = (c_1, c_2, \ldots, c_p) \in \mathbb{C}^{1 \times np}$ be a left eigenvector of A corresponding to the eigenvalue λ, i.e. $cA = \lambda c$ and $c \neq 0$. $c_1 \neq 0$ also follows from the proof of Lemma 5.1. From the Yule–Walker equations, it also follows (see (5.15)) that

$$\Gamma_p = A \Gamma_p A' + B \Sigma B'$$

and hence $c \Gamma_p c^* = c A \Gamma_p A' c^* + c B \Sigma B' c^* = |\lambda|^2 c \Gamma_p c^* + c_1 \Sigma c_1^*$ resp.

$$(1 - |\lambda|^2) c \Gamma_p c^* = c_1 \Sigma c_1^*.$$

That is, if Γ_p is positive semidefinite, then $|\lambda| \leq 1$ must hold, and in the case of $\Gamma_p > 0$ and $\Sigma = \Sigma_{1,p} > 0$, it follows that $|\lambda| < 1$. The condition $\Gamma_p > 0$ and $\Sigma = \Sigma_{1,p} > 0$ is equivalent to $\Gamma_{p+1} > 0$; see (2.11).

Now suppose that $\mathrm{rk}(\Gamma_p) = m < np$. We choose a special basis for the subspace

$$\mathrm{sp}\{x_{t-1}^p\} = \mathrm{sp}\{x_{1,t-1}, \ldots, x_{n,t-1}, x_{1,t-2}, \ldots, x_{n,t-2}, \ldots, x_{1,t-p}, \ldots, x_{n,t-p}\}$$

by selecting, in the natural order, the first linearly independent elements; see also, for example, Deistler et al. (2011). This means for the k-th components x_{ks} that we select the first m_k elements $x_{ks}, t - 1 \geq s \geq t - m_k$, and the remaining $x_{ks}, t - m_k > s$ are not selected. Of course, $m = m_1 + \cdots + m_n$. For this basis, we construct a selection matrix $S \in \mathbb{R}^{m \times np}$ (i.e. a matrix S, the entries of which are 0 or 1 and for which $SS' = I_m$ holds), such that Sx_t^p contains the basis elements and thus $S\Gamma_p S' > 0$. Now we choose a corresponding solution for the Yule–Walker equations

$$a = (a_1, \ldots, a_p) = (\gamma(1), \ldots, \gamma(p))S'(S\Gamma_p S')^{-1} S.$$

That is, the k-th column $\underline{a}(z)u_k$ of the associated polynomial $\underline{a}(z) = I_n - a_1 z - \cdots - a_p z^p$ is therefore a polynomial of degree $\delta_k \leq m_k$. Here, u_k denotes the k-th unit vector in $\mathbb{R}^n$. If c is a left eigenvector of A corresponding to the eigenvalue $\lambda \neq 0$, then, according to the proof of Lemma 5.1, the following holds:

$$c_1 \underline{a}(z)u_k = (1 - \lambda z)\underline{c}(z)u_k.$$

This means that the k-th element of $\underline{c}(z) = c_1 + \cdots + c_p z^{p-1}$ has degree less than or equal to $m_k - 1$. In other words, it holds that $c = cS'S$. Finally, the assertion $|\lambda| \leq 1$ follows from

$$(1 - |\lambda|^2)c\Gamma_p c^* = (1 - |\lambda|^2)cS'(S\Gamma_p S')Sc^* = c_1 \Sigma c_1^*$$

and $S\Gamma_p S' > 0$.   □

The above proposition shows that the Yule–Walker equations have particularly desirable properties in the case of $\Gamma_{p+1} > 0$; they are uniquely solvable and the solution corresponds to a stable AR system. The following two propositions now give sufficient conditions for ($\Gamma_k > 0 \; \forall k > 0$). The first proposition deals with the scalar case and holds for arbitrary positive semidefinite sequences $(\gamma(k))$, thus in particular also for the empirical autocovariance function $\hat{\gamma}(k)$.

Proposition 5.12 *For the scalar case, it follows from $\gamma(0) > 0$ and $\lim_{k \to \infty} \gamma(k) = 0$ that the Toeplitz matrices Γ_k (for all $k \geq 1$) are positive definite.*

Proof We perform a proof by contradiction and assume that $\Gamma_k > 0$ and $\det(\Gamma_{k+1}) = 0$ holds for a $k > 0$. Therefore, $\Gamma_k > 0$ and $\Sigma_{1,k} = 0$, and hence $x_{t+1} = a_1 x_t + \cdots + a_k x_{t+1-k}$ holds for the corresponding "prediction" coefficients $a_1, \ldots, a_k$. We consider the stacked process (x_t^k) and the corresponding companion matrix A. It holds that $x_{t+1}^k = Ax_t^k$ and thus $x_{t+m}^k = A^m x_t^k$ for all $m \geq 0$. The desired contradiction is now yielded by

$$\Gamma_k = \mathbf{Var}(x_{t+m}^k) = \mathbf{Var}(A^m x_t^k) = A^m \Gamma_k (A^m)' = (A^m \Gamma_k)\Gamma_k^{-1}(A^m \Gamma_k)'$$

and

$$A^m \Gamma_k = \mathbf{E}(A^m x_t^k)(x_t^k)' = \mathbf{E}x_{t+m}^k (x_t^k)' \overset{m \to \infty}{\longrightarrow} 0.$$

□

In the multivariate case, a similarly simple condition does not exist. However, the following proposition holds.

Proposition 5.13 *Let $\gamma(.)$ be the covariance function of a stationary process. If the variance of the innovations is positive definite, then it holds that $\Gamma_k > 0$ for all $k > 0$.*

Proof Let Σ_0 be the variance of the innovations. The assertion follows immediately from $\Sigma_{1,k-1} \geq \Sigma_0$ and the relation (2.12). □

5.5   The Unstable and Non-stationary Case

In this section, we briefly discuss the general case, i.e. we no longer assume that the stability condition is satisfied. Special emphasis is given to the "unit root case", where the AR polynomial has a zero at $z = 1$. This leads to integrated and cointegrated processes.

From Proposition 4.23, it follows immediately that an AR system has a unique stationary solution, if

$$\det \underline{a}(z) \neq 0 \ \forall |z| = 1.$$

This solution has a (in general two-sided) $MA(\infty)$ representation

$$x_t = \sum_{j=-\infty}^{\infty} k_j \epsilon_{t-j}$$

and is regular, as shown in Corollary 6.8 below. Hence, we see that only the zeros on the unit circle cause "problems". If $\det \underline{a}(z)$ has a zero on the unit circle, then no stationary solution exists or there are infinitely many stationary solutions (in the latter case, the variance $\Sigma = \mathbf{E}\epsilon_t\epsilon_t'$ must be singular). However, there is exactly one regular solution among the stationary solutions (if there are stationary solutions at all).

It is possible to show the following: If (x_t) is a regular AR process, then one can always find a stable AR system (that is, a system satisfying the stability condition) and white noise (ϵ_t), such that (x_t) is the (unique) stationary solution of this stable system. Consequently, one can assume (if one is interested only in regular processes) the stability condition without loss of generality.

In the following, we will briefly discuss non-stationary solutions of an AR system. However, we do not consider arbitrary non-stationary solutions, but only so-called *integrated processes*. Integrated processes play an important role, especially in econometrics, e.g. for modeling macroeconomic time series and financial data. Here, we focus on solutions on $\mathbb{N}$, i.e. we analyze the process (x_t) only for $t > 0$, that is, we require for the solution only, that it satisfies the difference Eq. (5.1) for $t > 0$.

It is clear that one can simply compute the solutions on $\mathbb{N}$ in a recursive way for arbitrary initial values $x_{1-p}, \ldots, x_0$. For the stacked process x_t^p, it follows, e.g. that

$$x_t^p = A x_{t-1}^p + B \epsilon_t = A^t x_0^p + \sum_{j=0}^{t-1} A^j B \epsilon_{t-j} \quad t > 0.$$

If the spectral radius $\varrho(A)$ of A is greater than one, then the variance $\mathbf{E} x_t x_t'$ in general diverges at a geometric rate $(\mathbf{E} x_t' x_t \geq c \varrho(A)^{2t})$. We exclude this exponentially unstable case; thus, it is required that $\varrho(A) \leq 1$. We also exclude complex eigenvalues on the unit circle as well as the eigenvalue -1. We therefore also exclude so-called seasonally integrated or cointegrated processes. That is, if $\lambda \in \mathbb{C}$ is an eigenvalue of A, then we require that

$$|\lambda| < 1 \text{ or } \lambda = 1.$$

The following condition on the zeros of $\det \underline{a}(z)$ is equivalent to this:

$$\det \underline{a}(z) = 0 \implies |z| > 1 \text{ or } z = 1.$$

$z = 1$ is also called unit root. The simplest case for an AR system with a unit root is the AR(1) system

$$x_t = I_n x_{t-1} + \epsilon_t$$

with the solution

$$x_t = x_0 + \sum_{j=1}^{t} \epsilon_j, \quad t > 0$$

which (for $x_0 = 0$) is a so-called random walk. The covariance matrix of x_t (for $x_0 = 0$) grows linearly in t:

$$\mathbf{E} x_t x_t' = t \mathbf{E} \epsilon_t \epsilon_t' = t \Sigma.$$

This means that the random walk is non-stationary (if $\Sigma = \mathbf{E} \epsilon_t \epsilon_t' \neq 0$); however, the first differences $x_t - x_{t-1} = \epsilon_t, t \geq 1$ are stationary. More generally, we now define the following.

Definition 5.14 (*processes integrated order of one*) A stochastic process $(x_t \mid t \geq t_0)$ is integrated of order one if $(x_t \mid t \geq t_0)$ is non-stationary, but $(x_t - x_{t-1} \mid t > t_0)$ is stationary. We often use the notation $(x_t) \sim I(1)$ for processes which are integrated of order one, and analogously $(u_t) \sim I(0)$ for a stationary process (u_t).

The simplest possibility to generate an I(1) process is to "integrate" a stationary process: Let $(u_t) \sim I(0)$ be a stationary process, then the process

$$x_t = x_0 + \sum_{j=1}^{t} u_j, \quad t > 0$$

is in general integrated of order $d = 1$. The following proposition gives a more precise characterization.

Proposition 5.15 (Beveridge–Nelson decomposition) *Let $(u_t) = \underline{k}(L)(\epsilon_t)$ be an n-dimensional, causal MA(∞) process, that is, $(\epsilon_t) \sim WN(\Sigma)$ is white noise and $\underline{k}(L) = \sum_{j \geq 0} k_j L^j$ is a causal filter.*
    Additionally, we require $\sum_{j \geq 0} j \|k_j\| < \infty$, then the process $x_t = x_0 + \sum_{j=1}^{t} u_j$, $t > 0$ has the representation

$$x_t = \sum_{j=1}^{t} \underline{k}(1)\epsilon_j + v_t + x_0^*, \ t > 0$$

where $x_0^* = x_0 - v_0$, $\underline{k}(1) = \sum_{j \geq 0} k_j$, $(v_t) = \tilde{\underline{k}}(L)(\epsilon_t)$ is a (causal) MA(∞) process and $\tilde{\underline{k}}(L) = \sum_{j \geq 0} \tilde{k}_j L^j$ is a (causal) l_1-filter determined by the identity $\underline{k}(z) = \underline{k}(1) + (1 - z)\tilde{\underline{k}}(z)$.

Proof The coefficients of $\tilde{\underline{k}}(z)$ are determined by a comparison of coefficients in the equation $\underline{k}(z) = \underline{k}(1) + (1 - z)\tilde{\underline{k}}(z)$:

$$k_0 = \underline{k}(1) + \tilde{k}_0 \qquad \Longrightarrow \qquad \tilde{k}_0 = -k_1 - k_2 - k_3 - \cdots$$
$$k_1 = \tilde{k}_1 - \tilde{k}_0 \qquad \Longrightarrow \qquad \tilde{k}_1 = -k_2 - k_3 - k_4 - \cdots$$
$$\vdots \qquad\qquad\qquad\qquad\qquad \vdots$$

One then obtains $\tilde{k}_j = -\sum_{l > j} k_l$ for $j \geq 0$. The coefficients $(\tilde{k}_j \mid j \geq 0)$ are absolutely summable, since

$$\sum_{j \geq 0} \|\tilde{k}_j\| \leq \sum_{j \geq 0} \sum_{l > j} \|k_l\| = \sum_{j > 0} j \|k_j\| < \infty.$$

Thus, the filter $\tilde{\underline{k}}(L)$ is an l_1-filter and $(v_t) = \tilde{\underline{k}}(L)(\epsilon_t)$ is an MA(∞) process. Finally, it follows from $(u_t) = \underline{k}(L)(\epsilon_t) = \underline{k}(1)(\epsilon_t) + \tilde{\underline{k}}(L)(1 - L)(\epsilon_t)$ that

$$u_t = \underline{k}(1)\epsilon_t + v_t - v_{t-1}$$

and hence the assertion

$$x_t = x_0 + \sum_{j=1}^{t} u_j = x_0 + \sum_{j=1}^{t} \underline{k}(1)\epsilon_j + v_t - v_0.$$

$\square$

In order to simplify the discussion of this result, we assume that the initial value x_0^* is uncorrelated to (ϵ_t), i.e. $\mathbf{E}x_0^*\epsilon_t = 0 \ \forall t \in \mathbb{Z}$. This also includes the case that x_0^* is deterministic.

If $\underline{k}(1)\Sigma\underline{k}(1)' > 0$, then for large t, the process (x_t) behaves essentially like a random walk, since the variance of the term $\sum_{j=1}^{t} \underline{k}(1)\epsilon_j$ grows linearly with t while the other terms have bounded variance. This means that for an asymptotic analysis of estimators, it is essentially sufficient to understand the behavior of these estimators for the "random walk case".

However, the process (x_t) is not integrated in all cases. If $\underline{k}(1)\Sigma\underline{k}(1)' = 0$, then the process (x_t) is stationary.

In the multivariate case $n > 1$, it may occur that

$$0 < q = \mathrm{rk}(\underline{k}(1)\Sigma\underline{k}(1)') < n.$$

If $\beta \in \mathbb{R}^{n \times 1}$ is contained in the (right) kernel of $\underline{k}(1)\Sigma\underline{k}(1)'$, then $\beta'\underline{k}(1)\epsilon_t = 0$ a.s. and hence

$$\beta'x_t = \beta'x_0^* + \sum_{j=1}^{t} \beta'\underline{k}(1)\epsilon_j + \beta'v_t = \beta'x_0^* + \beta'v_t.$$

That is, the process (x_t) is non-stationary, but there are certain linear combinations $(\beta'x_t)$ that are stationary.

Definition 5.16 An integrated process $(x_t) \sim I(1)$ is called *cointegrated*, if a vector $\beta \neq 0$ exists such that $(\beta'x_t) \sim I(0)$ is stationary. Such vectors are called *cointegration vectors* or *cointegrating vectors*. The dimension of the subspace spanned by all cointegrating vectors is called the cointegration rank of (x_t).

The cointegrating vectors β are often interpreted as long-run equilibrium relations between the variables $x_{1t}, \ldots, x_{nt}$. Since $(\beta'x_t)$ is stationary and regular, only short-term fluctuations around the equilibrium point $\beta'x_0^*$ are possible. By factorizing the variance $\underline{k}(1)\Sigma\underline{k}(1)' = BB'$, $B \in \mathbb{R}^{n \times q}$ and using a left inverse $B^{\dagger} \in \mathbb{R}^{q \times n}$ of B, we can also write (x_t) as

$$x_t = B\left(\sum_{j=1}^{t} \eta_j\right) + v_t + x_0^* \tag{5.20}$$

where $(\eta_t) \sim WN(I_q)$ is defined by $\eta_t = B^{\dagger}\epsilon_t$. The space spanned by the cointegration vectors is the (right) kernel of B' or of $\underline{k}(1)\Sigma\underline{k}(1)'$, respectively. The cointegration rank of (x_t) is hence equal to $r = n - q$. The representation above also shows that the process (x_t) is dominated by q uncorrelated random walk processes $(\sum_{j=1}^{t} \eta_{ij})$, $i = 1, \ldots, q$.

The structure of (x_t) is to a large extent determined by $V = \underline{k}(1)\Sigma\underline{k}(1)'$, the so-called *long-run variance* of (u_t). This long-run variance is, up to a factor 2π, equal to the spectral density of (u_t) at the point $\lambda = 0$:

$$V = \underline{k}(1)\Sigma\underline{k}(1)' = 2\pi(\sum_{j\geq 0} k_j e^{i\lambda 0})\frac{1}{2\pi}\Sigma(\sum_{j\geq 0} k_j e^{i\lambda 0})^* = 2\pi f_u(0).$$

Let us now return to the AR system (5.1) and impose the following conditions:

GR.1 $E\epsilon_t\epsilon_t' = \Sigma > 0$.
GR.2   If $\det \underline{a}(z) = 0$, then $z = 1$ or $|z| > 1$.
GR.3   The matrix $\Pi = -\underline{a}(1)$ has rank $0 < r = \mathrm{rk}(\Pi) < n$.

The matrix Π can hence be factorized as $\Pi = \alpha\beta', \alpha, \beta \in \mathbb{R}^{n\times r}$. One now constructs matrices $\bar{\alpha}, \bar{\beta} \in \mathbb{R}^{n\times r}, \alpha_\perp, \beta_\perp, \bar{\alpha}_\perp, \bar{\beta}_\perp \in \mathbb{R}^{n\times q}$, where $q = n - r$, such that

$$\begin{pmatrix} \alpha' \\ \alpha_\perp' \end{pmatrix} \begin{pmatrix} \bar{\alpha} & \alpha_\perp \end{pmatrix} = \begin{pmatrix} \beta' \\ \beta_\perp' \end{pmatrix} \begin{pmatrix} \bar{\beta} & \beta_\perp \end{pmatrix} = I_n.$$

For the polynomial matrix $\tilde{\underline{a}}(z) = I - \tilde{a}_1 z - \cdots - \tilde{a}_{p-1} z^{p-1}$ defined by the equation

$$\underline{a}(z) = -\Pi z + (1 - z)\tilde{\underline{a}}(z), \tag{5.21}$$

we now require that

GR.4   The matrix $(\alpha_\perp' \tilde{\underline{a}}(1)\beta_\perp) \in \mathbb{R}^{q\times q}$ has full rank q.

Proposition 5.17 (Granger's representation theorem) *Consider an AR system (5.1) satisfying the conditions GR.1–GR.4 above. Then*

(1) The rational matrix functions

$$\underline{k}(z) = (a(z)(1 - z)^{-1})^{-1} \text{ and } \tilde{\underline{k}}(z) = (1 - z)^{-1}(\underline{k}(z) - \underline{k}(1))$$

have no poles for $|z| \leq 1$. Therefore, they define two causal l_1-filters $\underline{k}(L)$ and $\tilde{\underline{k}}(L)$.
(2) The AR system has a solution of the form

$$x_t = \underline{k}(1)\sum_{j=1}^{t} \epsilon_j + v_t + x_0^* \tag{5.22}$$

where $(v_t) = \tilde{\underline{k}}(L)(\epsilon_t)$ and $\beta' x_0^* = 0$. It holds that

$$\underline{k}(1) = (\beta_\perp(\alpha_\perp' \tilde{\underline{a}}(1)\beta_\perp)^{-1}\alpha_\perp'). \tag{5.23}$$

(3) The processes $\Delta(x_t) = \underline{k}(L)(\epsilon_t)$ and $(\beta' x_t) = (\beta' v_t)$ are stationary. (Here, $\Delta = (1 - L)$ denotes the difference filter as usual.)

(4) The process (x_t) is cointegrated and the space of the cointegrating vectors is the column space of β'. The cointegration rank of (x_t) is equal to r.

Proof From the assumptions, it follows that

$$\underline{a}(z) \begin{bmatrix} \overline{\beta} & \beta_\perp \end{bmatrix} = \begin{bmatrix} \underline{a}(z)\overline{\beta} & \tilde{a}(z)\beta_\perp(1-z) \end{bmatrix} = \check{a}(z) \begin{bmatrix} I_r & 0 \\ 0 & I_q(1-z) \end{bmatrix}.$$

The polynomial matrix $\check{a}(z) = \begin{bmatrix} \underline{a}(z)\overline{\beta} & \tilde{a}(z)\beta_\perp \end{bmatrix}$ has only zeros outside the unit circle. For $z \neq 1$, $\det \underline{a}(z) = 0$ follows from $\det \check{a}(z) = 0$, and therefore, $|z| > 1$ must hold for the zeros of $\check{a}(z)$, according to (GR.2). The matrix $\check{a}(1)$ is non-singular, since the matrix

$$\begin{bmatrix} \overline{\alpha}' \\ \alpha'_\perp \end{bmatrix} \check{a}(1) = \begin{bmatrix} \overline{\alpha}' \\ \alpha'_\perp \end{bmatrix} \begin{bmatrix} -\alpha & \tilde{a}(1)\beta_\perp \end{bmatrix} = \begin{bmatrix} -I_r & \overline{\alpha}'\tilde{a}(1)\beta_\perp \\ 0 & \alpha'_\perp\tilde{a}(1)\beta_\perp \end{bmatrix}$$

is non-singular because of (GR.4). The process (x_t) is a solution of (5.1) if and only if the appropriately transformed process

$$\check{x}_t = \begin{bmatrix} \beta' x_t \\ \overline{\beta}'_\perp(x_t - x_{t-1}) \end{bmatrix}$$

is a solution of the AR system

$$\check{a}(L)(\check{x}_t) = (\epsilon_t). \tag{5.24}$$

We now assume that $(\check{x}_t)$ is the unique stationary solution of (5.24), i.e. we set

$$(\check{x}_t) = \check{a}^{-1}(L)(\epsilon_t)$$

where the inverse filter $\check{a}^{-1}(L)$ is determined as in Proposition 4.23. Hence, it follows that

$$\Delta(x_t) = \begin{bmatrix} \overline{\beta} & \beta_\perp \end{bmatrix} \begin{bmatrix} I_r\Delta & 0 \\ 0 & I_q \end{bmatrix} \check{a}^{-1}(L)(\epsilon_t) = \underline{k}(L)(\epsilon_t).$$

The transfer function of the filter $\underline{k}(L)$

$$(\overline{\beta}, \beta_\perp) \begin{pmatrix} I_r(1-z) & 0 \\ 0 & I_q \end{pmatrix} \left[\underline{a}(z)(\overline{\beta}, \beta_\perp) \begin{pmatrix} I_r & 0 \\ 0 & I_q(1-z)^{-1} \end{pmatrix} \right]^{-1} = \left(\underline{a}(z)(1-z)^{-1} \right)^{-1}$$

is a rational matrix function which has no poles in or on the unit circle. Therefore, the coefficients of $\underline{k}(L)$ decay geometrically and we can construct the Beveridge–Nelson decomposition for (x_t). From the above relations, the representation (5.23) for $\underline{k}(1)$ is also easy to derive. Moreover, the initial value x_0^* must satisfy the condition $\beta' x_0^* = 0$, so that $(\check{x}_t) = ((x_t'\beta, (x_t - x_{t-1})'\overline{\beta}_\perp)')$ is the stationary solution of (5.24) given above. $\qquad\square$

The proposition gives an explicit representation of a solution of the AR system. The proposition shows, in particular, how the space of cointegration vectors depends on the parameters of the system. The condition (GR.4) guarantees that no solutions exist with integration order $d > 1$.

By using Eq. (5.21), one obtains the so-called vector error correction model (VECM) representation

$$(x_t - x_{t-1}) = \alpha\beta' x_{t-1} + \tilde{a}_1(x_{t-1} - x_{t-2}) + \cdots \tilde{a}_{p-1}(x_{t+1-p} - x_{t-p}) + \epsilon_t.$$

The term $\alpha\beta' x_{t-1}$ can be interpreted as an error correction term that "drives back" the process to its long-run equilibrium $\beta' x_t = 0$. This form of the AR model is also commonly used to estimate the parameters and construct tests for the cointegration rank r.

References

T.W. Anderson, *The Statistical Analysis of Time Series* (Wiley, 1971)

M. Deistler, A. Filler, M. Funovits, AR systems and AR processes: the singular case. Commun. Inform. Syst. **11**(3), 225–236 (2011)

J. Durbin, The fitting of time series models. Rev. Inst. Int. Stat. **28**, 233–243 (1960)

R.F. Engle, C.W.J. Granger, Co-integration and error correction: representation, estimation, and testing. Econometrica, **55**(2), 251–276 (1987). ISSN 00129682. http://www.jstor.org/stable/1913236

E.J. Hannan, *Multiple Time Series* (Wiley, New York, 1970)

S. Johansen, *Likelihood-Based Inference in Cointegrated Vector Autoregressive Models* (Oxford University Press, 1995)

N. Levinson, The Wiener RMS error criterion in filter design and prediction. J. Math. Phys. **25**, 261–278 (1947)

H. Lütkepohl, *New Introduction to Multiple Time Series Analysis* (Springer, Berlin, 2005)

H.B. Mann, A. Wald, On the statistical treatment of linear stochastic difference equations. Econometrica **11**(3/4), 173–220 (1943)

P.C. Phillips, Time series regression with a unit root. Econometrica **55**(2), 277–301, März (1987). ISSN 00129682, 14680262. http://www.jstor.org/stable/1913237

G.U. Yule, On a method of investigating periodicities in disturbed series, with special reference to wolfer's sunspot numbers. Philos. Trans. R. Soc. Lond. A **226**, 267–298 (1927). Wiederabgedruckt in Stuart, Kendall (1971)

ARMA Systems and ARMA Processes

6

ARMA (AutoRegressive Moving Average) systems are of the form

$$x_t = a_1 x_{t-1} + \cdots + a_p x_{t-p} + \epsilon_t + b_1 \epsilon_{t-1} + \cdots + b_q \epsilon_{t-q} \tag{6.1}$$

where $a_j, b_j \in \mathbb{R}^{n \times n}$ are parameter matrices and $(\epsilon_t) \sim \mathrm{WN}(\Sigma)$ is white noise. *ARMA processes* are stationary solutions of ARMA systems.

In the first step, we describe the solutions of ARMA systems and the associated spectral densities. However, the focus is placed on the inverse question of how to obtain the underlying ARMA system, i.e. the ARMA parameters $a_j, j = 0, \ldots, p$, $b_j, j = 0, \ldots, q$ and $\Sigma = \mathbf{E} \epsilon_t \epsilon_t'$, from the spectral density. This is an important step on the way to estimating the ARMA parameters from a sample $x_1, \ldots, x_T$.

Regular ARMA processes are processes with rational spectral density; conversely, we will show that any process with rational spectral density is an ARMA process. AR and ARMA (and the equivalent state-space) models are the most important models for stationary processes. Any regular stationary process can be approximated (by appropriate choice of p and q) by an AR or ARMA process with arbitrary accuracy. For given "specification parameters" p and q, the parameter spaces of the corresponding classes of AR or ARMA systems are finite-dimensional. In this sense, we get a parametric estimation problem.

In comparing AR and ARMA modeling, the estimation of AR models is much simpler: There is no so-called identifiability problem in AR models, and the Yule–Walker equations, for example, give a linear system of equations for $a_1, \ldots, a_p$, which is not only simple to solve, but also yields consistent and asymptotically efficient estimators. In contrast, the estimation in the ARMA case is significantly more complex; there is an identifiability problem and important estimators like the maximum likelihood estimators are not available in explicit form, but have to be determined by numerical optimization. On the other hand, AR systems, as opposed

© Springer Nature Switzerland AG 2022
M. Deistler and W. Scherrer, *Time Series Models*, Lecture Notes in Statistics 224,
https://doi.org/10.1007/978-3-031-13213-1_6

to ARMA systems, are not closed to important operations such as marginalization.[1]
ARMA systems are also more flexible and often, one needs significantly more AR
than ARMA parameters for the approximation of a process with a given precision.

The book of Box and Jenkins (1976) has been a milestone in the modeling of
scalar time series by ARMA systems.[2,3] For the literature on multivariate ARMA
systems, we refer to Caines (1988), Hannan and Deistler (2012), Ljung (1999),
Lütkepohl (2005), Reinsel (1997), Scherrer and Deistler (2019) and Söderström and
Stoica (1989) as well to references given there.

6.1   ARMA Systems and Their Solutions

We consider ARMA systems of the form (6.1), always assuming that the *stability
condition*

$$\det(\underline{a}(z)) \neq 0, \quad \forall |z| \leq 1 \tag{6.2}$$

as well as the *minimum-phase condition*

$$\det(\underline{b}(z)) \neq 0, \quad \forall |z| < 1 \tag{6.3}$$

apply. Here, $\underline{a}(z) = I_n - a_1 z - \cdots - a_p z^p$ and $\underline{b}(z) = I_n + b_1 z + \cdots + b_q z^q$
denote the corresponding AR and MA polynomial matrices, respectively. Then the
unique stationary solution of (6.1) is of the form

$$(x_t) = \underline{a}^{-1}(L)\underline{b}(L)(\epsilon_t). \tag{6.4}$$

It follows directly from (6.4) and (4.10) that the spectral density of an ARMA process
is of the form

$$\underline{f}(z) = \frac{1}{2\pi}\underline{a}^{-1}(z)\underline{b}(z)\Sigma\underline{b}^*(z)\underline{a}^{-*}(z) \tag{6.5}$$

and hence *rational*. If (6.3) is replaced by

$$\det(\underline{b}(z)) \neq 0, \quad \forall |z| \leq 1, \tag{6.6}$$

one speaks of the *strict minimum-phase condition* or the *inverse stability condition*.
From the strict minimum-phase condition, it follows immediately that

$$(\epsilon_t) = \underline{b}^{-1}(L)\underline{a}(L)(x_t)$$

---

[1] Marginalization here refers to forming sub-processes from the original process.
[2] George E. P. Box (1919–2013), British-US statistician. Works on time series analysis, statistical
experimental design and Bayesian statistics.
[3] Gwilym M. Jenkins (1932–1982), British statistician and systems engineer. Known for the Box–
Jenkins method for the estimation of ARIMA models.

where $\underline{b}^{-1}(L)\underline{a}(L)$, as well as $\underline{a}^{-1}(L)\underline{b}(L)$, is causal. From this representation, we also obtain a so-called AR(∞) representation of the process

$$x_t = \sum_{j=1}^{\infty} \tilde{a}_j x_{t-j} + \epsilon_t.$$

These considerations show that, under the stability and strict minimum-phase condition, (6.4) is the Wold representation of the process (x_t). As we will show now, this also holds, if we replace (6.6) by the more general condition (6.3). The transfer function $\underline{b}^{-1}(e^{-i\lambda})\underline{a}(e^{-i\lambda})$ is also square integrable in the case of zeros of det $\underline{b}(z)$ on the unit circle with respect to $f(\lambda) = \underline{f}(e^{-i\lambda})$. Thus, in this case, $\epsilon_t = \Phi^{-1}\left(\underline{b}^{-1}(e^{-i\lambda})\underline{a}(e^{-i\lambda})e^{i\lambda t}\right)$ is also a linear transformation of the process (x_t), and hence $\epsilon_t \in \mathbb{H}(x)^n$ holds as well. Here, Φ is the isometry between the time and the frequency domain of the process, as defined in Proposition 3.14. In the following, we will rather sloppily write $(\epsilon_t) = \underline{b}^{-1}(L)\underline{a}(L)(x_t)$. However, to show that the ϵ_t's are innovations of (x_t), we have to show $\epsilon_t \in (\mathbb{H}_t(x))^n$. We write $\underline{b}^{-1}(z)$ as $\underline{b}^{-1}(z) = \underline{d}^{-1}(z)\underline{c}(z)$, where $\underline{d}(z)$ is a scalar polynomial, the zeros of which all lie on the unit circle, and $\underline{c}(z)$ is a rational matrix that has no poles inside or on the unit circle (which means that it is stable). The process $(y_t) = \underline{c}(L)\underline{a}(L)(x_t) = \underline{d}(L)(\epsilon_t)$ is an MA process. The Wold representation of the k-th component is therefore of the form $(y_{kt}) = \tilde{\underline{d}}(L)(\tilde{\epsilon}_{kt})$, where $\tilde{\underline{d}}(z)$ is a scalar polynomial. The spectral density of (y_{kt}) equals

$$\underline{d}(z)\sigma_{kk}^2\underline{d}^*(z) = \tilde{\underline{d}}(z)\tilde{\sigma}_{kk}^2\tilde{\underline{d}}^*(z)$$

where σ_{kk}^2 resp. $\tilde{\sigma}_{kk}^2$ denote the variances of ϵ_{kt} and $\tilde{\epsilon}_{kt}$. It follows from this, as can easily be seen (also compare the factorization of scalar spectra in the next section), with the normalization $\underline{d}(0) = \tilde{\underline{d}}(0) = 1$ that $\underline{d}(z) = \tilde{\underline{d}}(z)$ and $\sigma_{kk}^2 = \tilde{\sigma}_{kk}^2$. The covariance matrices of the vectors $(\epsilon_{kt}, y_{kt}, y_{k,t-1}, \ldots, y_{k,t-m})'$ and $(\tilde{\epsilon}_{kt}, y_{kt}, y_{k,t-1}, \ldots, y_{k,t-m})'$ are thus identical and it follows for the projection P_m onto sp$(y_{kt}, y_{k,t-1}, \ldots, y_{k,t-m})$ that $P_m \epsilon_{kt} = P_m \tilde{\epsilon}_{kt}$. Since $\tilde{\epsilon}_{kt} \in \mathbb{H}_t(y)$, it now follows that

$$\tilde{\epsilon}_{kt} = \underset{m\to\infty}{\text{l.i.m}} \, P_m \tilde{\epsilon}_{kt} = \underset{m\to\infty}{\text{l.i.m}} \, P_m \epsilon_{kt} = \epsilon_{kt}.$$

However, this means that $\epsilon_{kt} \in \mathbb{H}_t(y)$ and hence, as claimed, $\epsilon_{kt} \in \mathbb{H}_t(x)$, since $(y_t) = \underline{c}(L)\underline{a}(L)(x_t)$ and $\underline{c}(L)\underline{a}(L)$ is a causal filter. Thus we have shown the following.

Lemma 6.1 *Under the stability condition* (6.2) *and the minimum-phase condition* (6.3), (6.4) *is the Wold representation of* (x_t).

Thus, in particular, under these conditions ARMA processes are regular stationary processes. In the following, we will also show the converse, i.e. we will show that regular ARMA processes may be represented by stable and minimum-phase ARMA systems.

In the following exercises, we consider the ARMA process (x_t), which is generated by the stable ARMA system

$$x_t = a_1 x_{t-1} + \cdots + a_p x_{t-p} + \epsilon_t + b_1 \epsilon_{t-1} + \cdots + b_q \epsilon_{t-q}$$

with $(\epsilon_t) \sim \mathrm{WN}(\Sigma)$. We also assume without loss of generality that $p = q = m$. Let the causal MA(∞) representation of x_t be $x_t = \sum_{j \geq 0} k_j \epsilon_{t-j}$, i.e. $\underline{a}^{-1}(z)\underline{b}(z) = \underline{k}(z) = \sum_{j \geq 0} k_j z^j$.

Exercise 6.2 Show that the coefficients of the MA(∞) representation can be determined recursively by the following equations (see also (4.30)):

$$
\begin{aligned}
k_0 &= I \\
k_1 &= a_1 k_0 + b_1 \\
&\;\;\vdots \\
k_m &= a_1 k_{m-1} + \cdots + a_m k_0 + b_m \\
k_j &= a_1 k_{j-1} + \cdots + a_m k_{j-m} \quad \text{for } j > m.
\end{aligned}
\tag{6.7}
$$

Exercise 6.3 Show that the covariance function γ of (x_t) satisfies the following "generalized Yule–Walker equations":

$$\gamma(0) = a_1 \gamma(-1) + \cdots + a_m \gamma(-m) + \sum_{j=0}^{m} b_j \Sigma k_j'$$

$$\gamma(1) = a_1 \gamma(0) + \cdots + a_m \gamma(1-m) + \sum_{j=1}^{m} b_j \Sigma k_{j-1}'$$

$$\vdots$$

$$\gamma(m) = a_1 \gamma(m-1) + \cdots + a_m \gamma(0) + b_m \Sigma k_0'$$

$$\gamma(j) = a_1 \gamma(j-1) + \cdots + a_m \gamma(j-m) \quad \text{for } j > m.$$

Note: Show $\mathbf{E}\epsilon_s x_t' = 0$ for $s > t$ and $\mathbf{E}\epsilon_s x_t' = \Sigma k_{s-t}'$ for $s \leq t$ first and then proceed analogously as for the derivation of the Yule–Walker equations for AR processes.

The above equations can be used to determine the covariance function γ for given ARMA parameters. They also show that the MA(∞) coefficients (k_j) and the covariance function $\gamma(j)$ converge to zero at a geometric rate. (See also the corresponding exercises in the AR case.) From the equations for $j > m$, one can determine (for "generic" ARMA processes) the autoregressive parameters $a_1, \ldots, a_p$ from the covariance function.

Exercise 6.4 In the scalar case ($n = 1$), the following claim holds: Let (y_t) be the AR process $y_t = a_1 y_{t-1} + \cdots + a_m y_{t-m} + \epsilon_t$, then the ARMA process (x_t) may be written as

$$x_t = y_t + b_1 y_{t-1} + \cdots + b_m y_{t-m}.$$

This relation can also be used to determine the MA(∞) representation and the covariance function of (x_t).

Exercise 6.5 (*prediction of ARMA processes*) We now in addition assume that the minimum-phase condition holds and hence, (ϵ_t) is the innovation process of (x_t). Show the following representation (recursive in h) of the h-step ahead predictions from the *infinite* past

$$\hat{x}_{t,1} = a_1 x_t + \cdots + a_m x_{t+1-m} + b_1 \epsilon_t + \cdots + b_m \epsilon_{t+1-m}$$
$$\hat{x}_{t,2} = a_1 \hat{x}_{t,1} + a_2 x_t + \cdots + a_p x_{t+2-m} + b_2 \epsilon_t + \cdots + b_m \epsilon_{t+2-m}$$

$$\vdots$$

$$\hat{x}_{t,m} = a_1 \hat{x}_{t,m-1} + \cdots + a_{m-1} \hat{x}_{t,1} + a_m x_t + b_m \epsilon_t$$
$$\hat{x}_{t,h} = a_1 \hat{x}_{t,h-1} + \cdots + a_m \hat{x}_{t,h-m} \quad \text{for } h > m.$$

Here, $m = \max(p, q)$ and we set $a_j = 0$ for $j > p$ and $b_j = 0$ for $j > q$. To express the prediction as a function of the past x_t's, one uses (under the strict minimum-phase assumption) the representation

$$(\epsilon_t) = \underline{b}^{-1}(L)\underline{a}(L)(x_t) = \left(\sum_{j \geq 0} l_j x_{t-j} \right).$$

This also shows that infinitely many past values of (x_t) are relevant for prediction if $\underline{b}(z)$ is not a unimodular matrix.

In the next two sections, we will show that the class of rational spectra is exactly the class of spectra of ARMA processes, and we will describe the so-called factorization of such spectra, which first leads from the second moments (or from the spectrum) to the transfer function of the Wold decomposition and, in the second step, to the ARMA parameters.

6.2   The Factorization of Rational Spectra

We will first discuss the scalar case ($n = 1$). Now let $f(z)$ be an arbitrary one-dimensional rational spectral density. We want to show that f has a representation (6.5) satisfying (6.2) and (6.3).

According to Proposition 3.5, it holds that $f(\lambda) = \underline{f}(e^{-i\lambda})$ is integrable, $f(\lambda) \geq 0$ and $f(-\lambda) = f(\lambda)$. As a rational function, we write $\underline{f}$ as the quotient of two relatively prime polynomials $\underline{p}$ and $\underline{q}$:

$$\underline{f}(z) = \frac{\underline{p}(z)}{\underline{q}(z)}.$$

Here, $\underline{q}(e^{-i\lambda}) \neq 0 \ \forall \lambda$ must hold; otherwise, f would not be integrable. Using the fundamental theorem of algebra, we write in evident notation

$$\underline{f}(z) = cz^r \frac{\prod_{j=1}^{m}(z - z_j)}{\prod_{j=1}^{n}(z - v_j)}$$

where $z_j \neq 0$, $v_j \neq 0$, $|v_j| \neq 1$ and the zero sets $\{z_j \mid j = 1, \ldots, m\}$ and $\{v_j \mid j = 1, \ldots n\}$ are disjoint. Moreover, the zeros z_j with $|z_j| = 1$ must occur with even multiplicity, since it holds that $f(\lambda) \geq 0$.

From the conditions $\overline{f(\lambda)} = f(\lambda)$, $f(-\lambda) = f(\lambda)$ and the identity

$$(z - u) = (-zu)(z^{-1} - u^{-1}) \text{ for } z, u \neq 0, \tag{6.8}$$

it follows that

$$f(\lambda) = \overline{f(\lambda)} = \overline{c}e^{i\lambda r} \frac{\prod_{j=1}^{m}(e^{i\lambda} - \overline{z_j})}{\prod_{j=1}^{n}(e^{i\lambda} - \overline{v_j})} = c' e^{-i\lambda(n-m-r)} \frac{\prod_{j=1}^{m}(e^{-i\lambda} - \overline{z_j}^{-1})}{\prod_{j=1}^{n}(e^{-i\lambda} - \overline{v_j}^{-1})}$$

$$f(\lambda) = f(-\lambda) = ce^{i\lambda r} \frac{\prod_{j=1}^{m}(e^{i\lambda} - z_j)}{\prod_{j=1}^{n}(e^{i\lambda} - v_j)} = c'' e^{-i\lambda(n-m-r)} \frac{\prod_{j=1}^{m}(e^{-i\lambda} - z_j^{-1})}{\prod_{j=1}^{n}(e^{-i\lambda} - v_j^{-1})}.$$

The rational extension $\underline{f}(z)$ of $f(\lambda)$ is unique, and we see that if $z_j \neq 0$ is a zero (resp. $v_j \neq 0$ a pole) of $\underline{f}$, then this also holds for $\overline{z_j}$, z_j^{-1} and $\overline{z_j}^{-1}$ (resp. $\overline{v_j}$, v_j^{-1} and $\overline{v_j}^{-1}$). The orders m, n must be even; it also holds that $r = (n - m - r)$, i.e. $m = 2q$, $n = 2p$ and $r = p - q$ for $p, q \in \mathbb{N}_0$.

We now order the zeros and poles according to their absolute value, such that $|z_j| \geq 1$, $z_{2q+1-j} = \overline{z_j}^{-1}$ holds for $1 \leq j \leq q$ and $|v_j| > 1$, $v_{2q+1-j} = \overline{v_j}^{-1}$ for $1 \leq j \leq p$. Thus, we can write the spectral density $\underline{f}$ (with (6.8)) as

$$\underline{f}(z) = cz^r \frac{\prod_{j=1}^{q}(z - z_j) \prod_{j=1}^{q}(z - \overline{z_j}^{-1})}{\prod_{j=1}^{p}(z - v_j) \prod_{j=1}^{p}(z - \overline{v_j}^{-1})} = c' \frac{\prod_{j=1}^{q}(z - z_j) \prod_{j=1}^{q}(z^{-1} - \overline{z_j})}{\prod_{j=1}^{p}(z - v_j) \prod_{j=1}^{p}(z^{-1} - \overline{v_j})}$$

where $c' > 0$ holds due to $f(\lambda) \geq 0$. If we now write

$$\underline{a}(z) = 1 + a_1 z + \cdots + a_p z^p = \prod_{j=1}^{p} (z - v_j) \prod_{j=1}^{p} (-v_j)^{-1}$$

$$\underline{b}(z) = 1 + b_1 z + \cdots + b_q z^q = \prod_{j=1}^{q} (z - z_j) \prod_{j=1}^{q} (-z_j)^{-1}$$

$$\sigma^2 = 2\pi c' \prod_{j=1}^{q} |z_j|^2 \prod_{j=1}^{p} |v_j|^{-2}$$

then the desired factorization of $\underline{f}$ follows as

$$\underline{f}(z) = \frac{1}{2\pi} \frac{\underline{b}(z)}{\underline{a}(z)} \sigma^2 \frac{\underline{b}^*(z)}{\underline{a}^*(z)}, \quad \sigma^2 > 0. \qquad (6.9)$$

The polynomials $\underline{a}(z)$, $\underline{b}(z)$ have real coefficients; $a_p \neq 0$, $b_q \neq 0$ hold and the stability condition $\underline{a}(z) \neq 0 \, \forall |z| \leq 1$ as well as the minimum-phase condition $\underline{b}(z) \neq 0 \, \forall |z| < 1$ are satisfied.

We have therefore proved the following proposition.

Proposition 6.6 (spectral factorization in the scalar case) *Any one-dimensional rational spectral density $\underline{f}$ that is nonzero can be uniquely represented as*

$$\underline{f}(z) = \frac{1}{2\pi} \underline{k}(z) \sigma^2 \underline{k}^*(z),$$

where $\sigma^2 > 0$ and $\underline{k}(z) = \underline{a}^{-1}(z)\underline{b}(z)$ is a rational function with $\underline{k}(0) = 1$, which has no poles for $|z| \leq 1$ and no zeros for $|z| < 1$.

According to the construction given above, $\underline{a}(z)$ and $\underline{b}(z)$ are relatively prime and, with the normalization $a_0 = b_0 = 1$, uniquely determined by $\underline{k}(z)$ and thus by $\underline{f}(z)$.

Now we generalize the proposition on spectral factorization to $n > 1$. Here, the basic idea is to find a rational matrix $\underline{H}$ for given $\underline{f}$ (with rank n), such that $\underline{H}\underline{f}\underline{H}^*$ is diagonal and then perform the factorization described above for the scalar case. For this purpose, we first define a rational matrix $\underline{H}_1$, which has ones in the main diagonal, whose $(j, 1)$ elements are of the form $-\underline{f}_{j1}(z)\underline{f}_{11}^{-1}(z)$ and which otherwise consists of nothing but zeros. Then $\underline{H}_1\underline{f}$ has, with the exception of the $(1, 1)$ element, zeros in the first column and in $\underline{H}_1\underline{f}\underline{H}_1^*$, the first row and the first column are equal to zero with the exception of the $(1, 1)$ element. We iterate this procedure now— in the next step for the right, lower $(n - 1) \times (n - 1)$ submatrix of $\underline{H}_1\underline{f}\underline{H}_1^*$— until $\underline{H}\underline{f}\underline{H}^*$ is diagonal. It is easy to see that $\underline{H}$ is rational, lower triangular and it has ones along the main diagonal. In particular, $\underline{H}$ is invertible. The diagonal elements of $\underline{H}\,f\,\underline{H}^*$ are scalar rational spectral densities that can be factorized as discussed above. Hence, it follows that

$$\underline{H}\,f\,\underline{H}^* = \underline{D}\,\underline{D}^*$$

where $\underline{D}$ is a rational diagonal matrix whose diagonal elements have no zeros for $|z| < 1$ and no poles for $|z| \le 1$. We have thus found a factorization of the form

$$f = \underline{H}^{-1}\underline{D}\,\underline{D}^*\underline{H}^{-*}$$

for the spectral density f. Now let $\underline{c}$ be a least common multiple of the denominator polynomials of $\underline{H}^{-1}$. According to the result above, then there exists a polynomial $\underline{d}$ with $\underline{d}(z) \ne 0$ for $|z| < 1$ such that $\underline{c}\,\underline{c}^* = \underline{d}\,\underline{d}^*$. Hence, it holds that

$$f = \underbrace{\underline{c}\,\underline{d}^{-1}\underline{H}^{-1}\underline{D}}_{\underline{S}}\,\underbrace{\underline{D}^*\underline{H}^{-*}\underline{d}^{-*}\underline{c}^*}_{\underline{S}^*} = \underline{S}\,\underline{S}^*$$

and $\underline{S}$ has no poles for $|z| \le 1$. We can exclude poles on the unit circle due to the integrability of $f(\lambda) = f(e^{-i\lambda})$.

In the next step, we use the so-called Blaschke factors to mirror the zeros of $\underline{S}$, which lie inside the unit circle, to the outside of the circle. Let $z_0, |z_0| < 1$ be a zero of $\underline{S}$, i.e. $\mathrm{rk}(\underline{S}(z_0)) = n - r < n$. Then there exists a constant unitary matrix $Q \in \mathbb{C}^{n \times n}$ such that the last r columns of $\underline{S}(z_0)Q$ are equal to zero. We can therefore write

$$\underline{S}(z)Q = \underline{\tilde{S}}(z)\begin{pmatrix} I_{n-r} & 0 \\ 0 & I_r(z - z_0) \end{pmatrix}.$$

The *Blaschke matrix*

$$\underline{B}(z) = Q\begin{pmatrix} I_{n-r} & 0 \\ 0 & I_r\frac{1-\bar{z_0}z}{z-z_0} \end{pmatrix}$$

is now unitary on the unit circle, and thus it holds that $f = \underline{S}\,\underline{S}^* = (\underline{S}\,\underline{B})(\underline{B}^*\underline{S}^*)$. For the new spectral factor

$$\underline{S}\,\underline{B} = \underline{\tilde{S}}(z)\begin{pmatrix} I_{n-r} & 0 \\ 0 & I_r(1 - \bar{z_0}z) \end{pmatrix}$$

the multiplicity of the zero z_0 was reduced and correspondingly, a zero was generated at $\bar{z_0}^{-1}$ (or the multiplicity of this zero was increased). This procedure is now carried out for all zeros inside the unit circle and thus a stable and minimum-phase rational matrix $\underline{k}(z)$ is obtained with

$$f(z) = \underline{k}(z)\underline{k}^*(z).$$

One can also show that a spectral factor $\underline{k}$ can be chosen, where the power series coefficients are real matrices.

In the following, we use a slightly different factorization of the form $\underline{f}(z) = \frac{1}{2\pi}\underline{k}(z)\Sigma\underline{k}^*(z)$ where we use the normalization $\underline{k}(0) = I_n$ (which results from the condition $a_0 = b_0 = I_n$). Note that Σ is the covariance matrix of the innovations. This then leads to the uniqueness of the stable and minimum-phase factors $\underline{k}$ and the covariance matrix Σ for given $\underline{f}$.

Proposition 6.7 (spectral factorization in the multivariate case) *Any rational spectral density $\underline{f}$ that has μ-a.e. full rank can be uniquely represented as*

$$\underline{f}(z) = \frac{1}{2\pi}\underline{k}(z)\Sigma\underline{k}^*(z)$$

where $\Sigma > 0$ and $\underline{k}$ is a rational stable minimum-phase matrix (with real coefficients) with $\underline{k}(0) = I_n$.

The only missing part is the proof of uniqueness, which we will add a little later. We will show first that this factorization of the spectrum leads to an ARMA representation of the underlying process and that $\underline{k}(z)$ is the transfer function of the Wold representation and Σ is the covariance matrix of the innovations. The rational matrix $\underline{k}$ can be written as $\underline{k}(z) = \underline{a}^{-1}(z)\underline{b}(z)$, where $\underline{a}$ and $\underline{b}$ are polynomial matrices, and $\underline{a}(z) = d(z)I_n$ where $d(z)$ (with $d(0) = 1$) is the least common multiple of the denominator polynomials of the entries of $\underline{k}(z)$ where for each entry the numerator and denominator polynomials are assumed to be relatively prime. By construction, $\underline{a}$ satisfies the stability condition and $\underline{b}$ the minimum-phase condition. The transfer function $k^{-1}(\lambda) = \underline{k}^{-1}(e^{-i\lambda})$ is square integrable with respect to $f(\lambda) = \underline{f}(e^{-i\lambda})$ (i.e. $k^{-1} \in (\mathbb{H}_F(x))^n$), since

$$k^{-1}(\lambda)f(\lambda)k^{-*}(\lambda) = \frac{1}{2\pi}\Sigma.$$

Hence, $(\epsilon_t) = \underline{k}^{-1}(\mathrm{L})(x_t) = \underline{b}^{-1}(\mathrm{L})\underline{a}(\mathrm{L})(x_t)$ is a valid transformation of the process (x_t) even if $\underline{k}$ has zeros on the unit circle and $\underline{k}^{-1}(\mathrm{L})$ is therefore not an l_1-filter. We also see that (ϵ_t) is white noise with variance $\mathbf{E}\epsilon_t\epsilon_t' = \Sigma$. According to this construction,

$$\underline{a}(\mathrm{L})(x_t) = \underline{b}(\mathrm{L})(\epsilon_t),$$

which means that (x_t) is an ARMA process. As shown at the end of Sect. 6.1, the ϵ_t's are innovations of (x_t) and $(x_t) = \underline{k}(\mathrm{L})(\epsilon_t) = \underline{a}^{-1}(\mathrm{L})\underline{b}(\mathrm{L})(\epsilon_t)$ is the Wold representation of (x_t).

With the help of this observation, the uniqueness of the factorization in Proposition 6.7 now follows immediately. For each such factorization $2\pi\underline{f} = \underline{k}\Sigma\underline{k}^*$, $\underline{k}$ is the transfer function of the Wold representation and Σ is the covariance matrix of the innovations. Since the Wold representation (for $\Sigma > 0$) is unique, it also follows that the factorization is unique.

Corollary 6.8 *A process with a rational spectral density that has μ-a.e. full rank is regular. The transfer function corresponding to the Wold representation is rational, stable and minimum-phase.*

Proof We consider the factorization $2\pi \underline{f} = \underline{k}\Sigma\underline{k}^*$ of the spectral density according to Proposition 6.7. Since $\underline{k}$ is rational, stable and minimum-phase, $\underline{k}$ is hence the transfer function of the Wold representation. $\qquad\square$

Example (Szegő formula)

The spectral factorization is also a key to the proof of the Szegő theorem 3.20. Here, we consider the simple case of a scalar process $(x_t) = \underline{k}(L)(\epsilon_t)$ with a rational, stable and strictly minimum-phase transfer function $\underline{k}$. The function $\log \underline{k}(z)$ is holomorphic on an open disk with the radius $\rho > 1$ and thus it follows with $2\pi f(\lambda) = \sigma^2 \underline{k}(e^{-i\lambda})\underline{k}(e^{i\lambda})$, $\underline{k}(0) = 1$ and Cauchy's integral formula:

$$\int_{-\pi}^{\pi} \log(2\pi f(\lambda))d\lambda = \int_{-\pi}^{\pi} \log \sigma^2 d\lambda + \int_{-\pi}^{\pi} \log(\underline{k}(e^{-i\lambda}))d\lambda + \int_{-\pi}^{\pi} \log(\underline{k}(e^{i\lambda}))d\lambda$$

$$= 2\pi \log \sigma^2 - 2\pi \log(\underline{k}(0)) + 2\pi \log(\underline{k}(0)) = 2\pi \log \sigma^2.$$

Example (non-stable AR system)

At the end of this section, we will briefly consider the case of an AR system where the stability condition is not necessarily satisfied. Let $(x_t) = \underline{\tilde{a}}^{-1}(L)(\tilde{\epsilon}_t)$ be the unique stationary solution of the AR system

$$x_t = \tilde{a}_1 x_{t-1} + \cdots + \tilde{a}_p x_{t-p} + \tilde{\epsilon}_t$$

with white noise $(\tilde{\epsilon}_t) \sim \text{WN}(\tilde{\Sigma})$, $\tilde{\Sigma} > 0$ and $\det \underline{\tilde{a}}(z) \neq 0$, $\forall |z| = 1$. The spectral density $\underline{f} = \frac{1}{2\pi}\underline{\tilde{a}}^{-1}(z)\tilde{\Sigma}\underline{\tilde{a}}^{-*}(z)$ can be factorized according to Proposition 6.7:

$$\underline{f}(z) = \frac{1}{2\pi}\underline{k}(z)\Sigma\underline{k}^*(z).$$

The inverse spectral density

$$\underline{f}^{-1}(z) = 2\pi\underline{\tilde{a}}^*(z)\tilde{\Sigma}^{-1}\underline{\tilde{a}}(z) = 2\pi\underline{k}^{-*}(z)\Sigma^{-1}\underline{k}^{-1}(z)$$

has no (finite) poles and $\underline{a}(z) := \underline{k}^{-1}(z)$ must hence be a polynomial matrix. The stability condition for $\underline{a}$ follows from the minimum-phase condition for $\underline{k}$ and the observation $\underline{f}(z) > 0$ for all $|z| = 1$. This way we have constructed an AR system for (x_t) that satisfies the stability condition.

Exercise 6.9 Determine the spectral factorization of

$$f(\lambda) = \frac{1}{2\pi} \frac{5.05 - \cos(\lambda)}{1.2 + 0.56\cos(\lambda) + 0.8\cos(2\lambda)}.$$

Note: First, determine the rational extension $f(z)$ of f. For this purpose, replace $\cos(k\lambda)$ by $\frac{1}{2}(e^{-i\lambda k} + e^{i\lambda k})$ and then $e^{-i\lambda k}$ by z^k.

Exercise 6.10 Show that the class of regular ARMA processes is closed with respect to the following operations. Here, (x_t) always denotes a regular, n-dimensional ARMA process. Note: It suffices to show that the constructed process (y_t) has a rational spectral density.

- Marginalization: $(y_t = Cx_t)$ for $C \in \mathbb{R}^{m \times n}$ is a regular ARMA process. Therefore, in particular, all component processes (x_{kt}) are regular ARMA processes.
- Summation: If (x_t) and (z_t) are two (n-dimensional) uncorrelated (i.e. $\mathbf{E}x_t z'_s = 0$ $\forall t, s \in \mathbb{Z}$) regular ARMA processes, then the summation $(y_t = x_t + z_t)$ is also a regular ARMA process.
- Rational filters: If $\underline{k}(z)$ is a rational transfer function that has no poles on the unit circle, then $(y_t) = \underline{k}(\mathrm{L})(x_t)$ is also a regular ARMA process.
- Sampling: The process $(y_s = x_{\Delta s})$ for $\Delta \in \mathbb{N}$ is a regular ARMA process. Note: See Exercise 3.25 (on sampling and aliasing).
- Aggregation: The process $(y_s = \sum_{i=1}^{\Delta} x_{\Delta s+i})$ for $\Delta \in \mathbb{N}$ is a regular ARMA process.

6.3   From the Wold Representation to ARMA Parameters: Observational Equivalence and Identifiability

A fundamental problem in the estimation of ARMA systems is the fact that ARMA parameters in general are not uniquely determined even for given population second moments (i.e. given spectral density). This a so-called *identifiability problem*. As shown in the previous section, the transfer function $\underline{k} = \underline{a}^{-1}\underline{b}$ (with the stability condition, the minimum-phase condition and the normalization $\underline{k}(0) = I_n$) is uniquely determined. However, the polynomial matrices $\underline{a}$ and $\underline{b}$ are not yet determined by this transfer function. It can be seen immediately that the pair $\tilde{\underline{a}} = \underline{c}\,\underline{a}$ and $\tilde{\underline{b}} = \underline{c}\,\underline{b}$, where $\underline{c}$ is an arbitrary non-singular polynomial matrix with $\underline{c}(0) = I_n$, yields the same transfer function $\underline{k} = \tilde{\underline{a}}^{-1}\tilde{\underline{b}} = \underline{a}^{-1}\underline{b}$. If $\underline{c}$ has no zeros inside or on the unit circle, then $\tilde{\underline{a}} = \underline{c}\,\underline{a}$ and $\tilde{\underline{b}} = \underline{c}\,\underline{b}$ satisfy the stability condition and the minimum-phase condition if $\underline{a}$ and $\underline{b}$ satisfy these conditions. In the scalar case, the order of the polynomials $\tilde{\underline{a}} = \underline{c}\,\underline{a}$ and $\tilde{\underline{b}} = \underline{c}\,\underline{b}$ is always strictly greater than the order of the polynomials $\underline{a}$ and $\underline{b}$, if $\underline{c}$ is not a constant. Therefore, in the scalar case, one obtains a unique decomposition $\underline{k} = \underline{a}^{-1}\underline{b}$, if one restricts oneself to relatively prime polynomials (resp. polynomials with minimum order). This no longer holds in the multivariate case; i.e. even the requirement of "minimum order" does not suffice here to uniquely determine $\underline{a}$ and $\underline{b}$. In a slightly more formal way, we can say that the mapping $(\underline{a}, \underline{b}) \longmapsto \underline{k} = \underline{a}^{-1}\underline{b}$ is not injective without additional restrictions on the domain of definition and hence $(\underline{a}, \underline{b})$ are not identifiable.

In this section, we consider model classes of ARMA systems: We assume that p and q are given and that the assumptions (6.2) and (6.3) as well as $a_0 = b_0 = I_n$ hold. This model class is described by the parameter space

$$\Theta_{p,q,V} = \Theta_{p,q} \times \Theta_V$$
$$\Theta_{p,q} = \big\{(a_1, \ldots, a_p, b_1, \ldots, b_q) \in \mathbb{R}^{n \times n(p+q)} \mid \det \underline{a}(z) \neq 0 \, \forall |z| \leq 1,$$
$$\det \underline{b}(z) \neq 0 \, \forall |z| < 1\big\}$$
$$\Theta_V = \big\{\Sigma \in \mathbb{R}^{n \times n} \mid \Sigma = \Sigma', \Sigma > 0\big\}.$$

The elements of $\Theta_{p,q,V}$ are called "real-valued parameters" in contrast to the integer "specification parameters" p and q. $\Theta_{p,q}$ is the corresponding space of system parameters. $\Theta_{p,q}$ is "thick" in $\mathbb{R}^{n \times (n(p+q))}$ in the sense that it contains a nontrivial open set, as can be shown. The set Θ_V of the innovation variances can be embedded in the $\mathbb{R}^{n(n+1)/2}$.

Here, the question of identifiability concerns the extent to which the parameters in $\Theta_{p,q,V}$ are uniquely determined from the corresponding spectrum $f(z)$. Due to Proposition (6.7), the innovation variance Σ and the stable and minimum-phase spectral factor $\underline{k}$ are always uniquely determined from the spectral density f; the question hence is whether the system parameters are uniquely determined from the transfer function $\underline{k}$, i.e. whether the representation $\underline{k} = \underline{a}^{-1}\underline{b}$ is unique. Clearly, such uniqueness is a reasonable requirement for parameter estimation. Therefore, we now consider the mapping π, defined on $\Theta_{p,q}$, which assigns the transfer function to the parameters, i.e. $\pi(a_1, \ldots, b_q) = \underline{k}(z) = \underline{a}^{-1}(z)\underline{b}(z)$.

Definition 6.11 For arbitrary $\underline{k} \in \pi(\Theta_{p,q})$, $\pi^{-1}(\underline{k})$ is called the *class of all observationally equivalent parameters* in $\Theta_{p,q}$ corresponding to $\underline{k}$. A subset $\Theta \subset \Theta_{p,q}$ is called *identifiable,* if the restriction of π to Θ is injective. This implies that for a transfer function $\underline{k} \in \pi(\Theta)$, the intersection of the equivalence class $(\pi^{-1}(\underline{k}))$ with Θ contains one element.

First, let us consider the scalar case again, and here the special case $p = q = 1$. The corresponding parameter space $\Theta_{1,1} = \{(a, b) \mid |a| < 1, |b| \leq 1\}$ is shown in Fig. 6.1.

We can see immediately that, whenever $(1 - az)$ and $(1 + bz)$ are relatively prime, i.e. $b \neq -a$, the parameters a, b (and hence also σ^2) are uniquely determined from $\underline{k}(z)$ and that this does not hold for the case $b = -a$. One should therefore exclude systems

$$x_t - ax_{t-1} = \epsilon_t - a\epsilon_{t-1},$$

all of which correspond to the transfer function $\underline{k}(z) = 1$. This transfer function corresponds to an "ARMA(0,0)" system.

Fig. 6.1 parameter space
$\Theta_{1,1}$ for $n = 1$

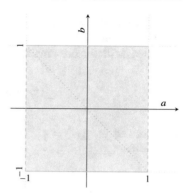

We have a similar picture in the general scalar case. Identifiability is obtained by additionally requiring relative primeness of $\underline{a}(z)$ and $\underline{b}(z)$, and the nontrivial equivalence classes in $\Theta_{p,q}$ correspond to pairs of polynomials of the form $(\underline{a}(z), \underline{b}(z)) = t(z)(\tilde{\underline{a}}(z), \tilde{\underline{b}}(z))$.

In the multivariate case, the analysis is more involved and one has to proceed as follows. We say that the pair $(\underline{a}(z), \underline{b}(z))$ of polynomial matrices has a *common (polynomial) left divisor* $t(z)$ if polynomial matrices $(\tilde{\underline{a}}(z), \tilde{\underline{b}}(z))$ exist, such that

$$(\underline{a}(z), \underline{b}(z)) = t(z)(\tilde{\underline{a}}(z), \tilde{\underline{b}}(z)). \tag{6.10}$$

$(\underline{a}(z), \underline{b}(z))$ are called *relatively left prime* if all common left divisors are *unimodular*, i.e. $\det t(z) \equiv \text{const} \neq 0$ holds for all common left divisors $t(z)$.

Lemma 6.12 (left prime polynomial matrices) *A pair* $(\underline{a}(z), \underline{b}(z))$ *is relatively left prime if and only if*

$$\text{rk}((\underline{a}(z), \underline{b}(z)) = n \quad \forall z \in \mathbb{C} \tag{6.11}$$

where n is the number of rows of $(\underline{a}(z), \underline{b}(z))$.

Proof If $(\underline{a}(z), \underline{b}(z))$ were not relatively left prime, a non-unimodular polynomial matrix would exist in (6.10), say $t(z)$. The determinant $\det t(z)$ would have a zero z_0 and thus

$$\text{rk}((\underline{a}(z_0), \underline{b}(z_0)) = \text{rk}(t(z_0)(\tilde{\underline{a}}(z_0), \tilde{\underline{b}}(z_0)) < n.$$

Conversely, if $\text{rk}((\underline{a}(z_0), \underline{b}(z_0)) < n$ holds for a $z_0 \in \mathbb{C}$, then there exists a constant non-singular matrix $C \in \mathbb{C}^{n \times n}$, such that the first row of $C(\underline{a}(z_0), \underline{b}(z_0))$ has only zeros. However, then

$$t(z) = C^{-1} \begin{pmatrix} z - z_0 & 0 \\ 0 & I_{n-1} \end{pmatrix}$$

is a non-unimodular left divisor of $(\underline{a}(z), \underline{b}(z))$.   □

Using the Smith–McMillan form (4.22), one can now construct a corresponding ARMA system $(\underline{a}(z), \underline{b}(z))$ from $\underline{k}(z)$ as follows: Let $\underline{k}(z) = \underline{u}(z)\underline{\Lambda}(z)\underline{v}(z)$ be the Smith–McMillan representation of $\underline{k}(z)$. The matrix

$$\underline{\Lambda}(z) = \text{diag}(\frac{p_{11}(z)}{q_{11}(z)}, \ldots, \frac{p_{nn}(z)}{q_{nn}(z)})$$

is factorized as $\underline{\Lambda}(z) = q^{-1}(z)p(z)$, where $q(z) = \text{diag}(q_{11}(z), \ldots, q_{nn}(z))$ and $p(z) = \text{diag}(p_{11}(z), \ldots, p_{nn}(z))$ are two diagonal polynomial matrices. Then,

$$\underline{a}(z) = q(z)\underline{u}^{-1}(z), \quad \underline{b}(z) = p(z)\underline{v}(z) \tag{6.12}$$

defines an ARMA system, corresponding to $\underline{k}(z)$, where the conditions on the zeros and poles of $\underline{k}(z)$ (see proposition on spectral factorization) guarantee the minimum-phase condition for $\underline{b}(z)$ and the stability condition for $\underline{a}(z)$. (The normalization of $a_0 = I_n$ and $b_0 = I_n$ is not required here, but it is easy to obtain.) Note that $\text{rk}(q(z), p(z)) = n \ \forall z \in \mathbb{C}$ follows from the properties of the polynomials $p_{ii}(z)$ and $q_{ii}(z)$ given in the proposition on the Smith–McMillan form. Hence, the pair $(q(z), p(z))$ is relatively left prime according to the above lemma. From

$$(q(z)\underline{u}^{-1}(z), p(z)\underline{v}(z)) = (q(z), p(z)) \begin{pmatrix} \underline{u}^{-1}(z) & 0 \\ 0 & \underline{v}(z) \end{pmatrix}$$

it follows with the unimodularity of $\underline{u}^{-1}(z)$ and $\underline{v}(z)$ that the left side of the above equation has a full rank for all $z \in \mathbb{C}$ as well. Hence, the pair $(\underline{a}(z), \underline{b}(z))$, defined in (6.12), is relatively left prime. Thus we have shown the following.

Proposition 6.13 *For every stable and minimum-phase rational transfer function $\underline{k}(z)$ (i.e. the poles of $\underline{k}(z)$ are all outside the unit circle and the zeroes are outside or on the unit circle), $\underline{k}(0) = I_n$, there exist a corresponding ARMA system $(\underline{a}(z), \underline{b}(z))$ satisfying the stability condition, the minimum-phase condition, the pair is relatively left prime and $\underline{a}(0) = \underline{b}(0) = I_n$ holds.*

The following proposition holds.

Proposition 6.14 *Two relatively left prime ARMA systems $(\underline{a}(z), \underline{b}(z))$ and $(\tilde{\underline{a}}(z), \tilde{\underline{b}}(z))$ are observationally equivalent if and only if a unimodular polynomial matrix $\underline{u}(z)$ with $\underline{u}(0) = I_n$ exists, such that*

$$(\tilde{\underline{a}}(z), \tilde{\underline{b}}(z)) = \underline{u}(z)(\underline{a}(z), \underline{b}(z)). \tag{6.13}$$

Proof One direction is evident, since

$$\tilde{\underline{a}}^{-1}(z)\tilde{\underline{b}}(z) = \underline{a}^{-1}(z)\underline{u}^{-1}(z)\underline{u}(z)\underline{b}(z) = \underline{a}^{-1}(z)\underline{b}(z).$$

Conversely, it follows from $\tilde{\underline{a}}^{-1}(z)\tilde{\underline{b}}(z) = \underline{a}^{-1}(z)\underline{b}(z)$ that

$$(\tilde{\underline{a}}(z), \tilde{\underline{b}}(z)) = \underbrace{(\tilde{\underline{a}}(z)\underline{a}^{-1}(z))}_{t(z)}(\underline{a}(z), \underline{b}(z)),$$

where $t(z)$ is a rational matrix. If $t(z)$ had poles, then due to the left primeness of $(\underline{a}(z), \underline{b}(z))$, $(\tilde{\underline{a}}(z), \tilde{\underline{b}}(z))$ would also have poles and it would not be a polynomial matrix. Thus, $t(z)$ must be a polynomial matrix. However, the relative left primeness of $(\tilde{\underline{a}}(z), \tilde{\underline{b}}(z))$ then implies that $t(z)$ is unimodular.   □

It follows immediately from the construction of $(\underline{a}(z), \underline{b}(z))$ from the Smith–McMillan form and the above proposition that for relatively left prime ARMA systems $(\underline{a}(z), \underline{b}(z))$, the poles of $\underline{a}^{-1}(z)\underline{b}(z)$ are equal to the zeros of $\underline{a}(z)$ and the zeros of $\underline{a}^{-1}(z)\underline{b}(z)$ are equal to the zeros of $\underline{b}(z)$.

Proposition 6.15 *The parameter space*

$$\tilde{\Theta}_{p,q} = \left\{ (a_1, \ldots, a_p, b_1, \ldots, b_q) \in \Theta_{p,q} \mid (\underline{a}(z), \underline{b}(z)) \text{ is relatively left prime,} \atop \mathrm{rk}(a_p, b_q) = n \right\}$$

is identifiable.

Proof According to the above proposition, between two observationally equivalent, left coprime systems, the relation (6.13) holds, with a unimodular $\underline{u}(z)$. Now $a_0 = b_0 = I_n$ holds for all systems corresponding to $\Theta_{p,q}$, so $\underline{u}(0) = I_n$ must also hold according to (6.13). If the degree of $\underline{u}(z)$ were greater than zero, it would follow from (6.13) (and the condition $\mathrm{rk}(a_p, b_q) = n$) that the degree of $\tilde{\underline{a}}(z)$ would be greater than p or the degree of $\tilde{\underline{b}}(z)$ would be greater than q. In other words, $\underline{u}(z)$ must be the unit matrix, and two arbitrary observationally equivalent systems corresponding to $\tilde{\Theta}_{p,q}$ are equal.   □

Exercise 6.16 Let (x_t) be the ARMA(1,0) process

$$x_t + \begin{pmatrix} \alpha & \alpha\beta \\ -\alpha\beta^{-1} & -\alpha \end{pmatrix} x_{t-1} = \epsilon_t$$

with $\alpha, \beta \in \mathbb{R}$, $\alpha, \beta \neq 0$. Show that (x_t) is also an ARMA(0,1) process, i.e. find a representation for the process of the form

$$x_t = \epsilon_t + b_1 \epsilon_{t-1}$$

with $b_1 \in \mathbb{R}^{2\times 2}$. Why is this example not a contradiction to the proposition on the identifiability of $\tilde{\Theta}_{p,q}$?

Exercise 6.17 Let $(\underline{a}(z), \underline{b}(z))$, $\det(\underline{a}(z)) \not\equiv 0$ be a pair of polynomial matrices of the order p and q. Show that by "reducing" the non-unimodular left divisors, one can construct a left prime pair $(\tilde{\underline{a}}(z), \tilde{\underline{b}}(z))$ of the order $\tilde{p}$ and $\tilde{q}$, such that $\underline{a}^{-1}(z)\underline{b}(z) = \tilde{\underline{a}}^{-1}(z)\tilde{\underline{b}}(z)$ and $\tilde{p} \le p, \tilde{q} \le q$.

In the scalar case, the order of the polynomials $\underline{a}$, $\underline{b}$ is minimal if and only if they are coprime to each other. The situation is more complicated in the multivariate case. However, this exercise shows that there are always left prime pairs among the observationally equivalent polynomial matrices with minimal order.

As can be shown, the parameter spaces $\tilde{\Theta}_{p,q}$ have the disadvantage that not all ARMA systems can be represented in such parameter spaces by way of an appropriate choice of p and q. However, through the suitable specification of the row or column degrees in $(\underline{a}(z), \underline{b}(z))$, any ARMA system can be embedded in an identifiable parameter space (see Hannan (1971) and Hannan and Deistler (2012, Chap. 2)).

The next proposition shows that $\Theta_{p,q} - \tilde{\Theta}_{p,q}$ is (in $\Theta_{p,q}$) a "thin" set:

Proposition 6.18 $\tilde{\Theta}_{p,q}$ *contains a subset which is open and dense in* $\Theta_{p,q}$.

Proof In the following, we will identify $\mathbb{R}^{n \times n(p+q)}$, the set of real matrices of dimension $n \times n(p+q)$, with the Euclidean space $\mathbb{R}^{n^2(p+q)}$ and correspondingly sets like $\Theta_{p,q}$ are considered as subsets of $\mathbb{R}^{n^2(p+q)}$. It is an elementary result that any polynomial function of several variables (i.e. a polynomial function $g : \mathbb{R}^m \longrightarrow \mathbb{R}, m \ge 1$), that is not identical zero, is nonzero on an open and dense subset of $\mathbb{R}^m$. Therefore, the sets

$$\Theta_a := \{(a_1, \ldots, b_q) \in \mathbb{R}^{n \times n(p+q)} \mid \det(a_p) \ne 0\}$$

$$\Theta_b := \{(a_1, \ldots, b_q) \in \mathbb{R}^{n \times n(p+q)} \mid \det(b_q) \ne 0\}$$

$$\Theta_S := \{(a_1, \ldots, b_q) \in \mathbb{R}^{n \times n(p+q)} \mid \det(\mathrm{Syl}(\det(\underline{a}(z), \det(\underline{b}(z)))) \ne 0\}$$

are open and dense in $\mathbb{R}^{n^2(p+q)}$. Here, $\mathrm{Syl}(\det(\underline{a}(z), \det(\underline{b}(z)))$ denotes the Sylvester matrix of the two scalar polynomials $\det \underline{a}(z)$ and $\det \underline{b}(z)$. However, note that this Sylvester matrix is constructed under the assumption that the polynomials have degrees np and nq, respectively. Clearly, the intersection of these three sets is again open and dense in $\mathbb{R}^{n^2(p+q)}$ and, above all, on this intersection $\det \underline{a}(z)$ and $\det \underline{b}(z)$ have no common divisor, since the Sylvester matrix has full rank. This implies that $(\underline{a}(z), \underline{b}(z))$ is relatively left prime, since otherwise there would be a common zero for $\det \underline{a}(z)$ and $\det \underline{b}(z)$ (see Kailath (1980, Sect. 2.4.4)). Finally, note that the set of stable and *strict* minimum-phase ARMA systems

$$\Theta_m := \{(a_1, \ldots, b_q) \in \mathbb{R}^{n \times n(p+q)} \mid \det(\underline{a}(z)) \ne 0 \, \forall |z| \le 1, \det(\underline{b}(z)) \ne 0 \, \forall |z| \le 1\}$$

is open in $\mathbb{R}^{n^2(p+q)}$ and dense in $\Theta_{p,q}$. In order to show "openness", we show that the set of polynomial matrices $\underline{a}$ which have a zero on or within the unit circle is closed.

Consider a sequence of such matrices $\underline{a}^{(k)}$ which converge to a matrix $\underline{a}^{(0)}$. Each of the matrices $\underline{a}^{(k)}$ has a zero, z_k say, with $|z^k| \leq 1$ and $\underline{a}^{(k)}(z^k) = 0$. Then there exists a convergent sub-sequence, which will be again indexed by k, such that $z^k \to z^0$. Clearly, $|z^0| \leq 1$ holds and $0 = \underline{a}^{(k)}(z^k) \to \underline{a}^{(0)}(z^0)$ implies that the set of matrices $(a_1, \ldots, b_q)$ where $\underline{a}(z)$ satisfies the stability assumption is open in $\mathbb{R}^{n^2(p+q)}$. The same argument shows that the set of matrices $(a_1, \ldots, b_q)$ where $\underline{b}(z)$ satisfies the strict minimum-phase assumption is open.

In order to show that Θ_m is dense in $\Theta_{p,q}$, we consider a polynomial matrix $\underline{b}(z) = b_0 + \cdots + b_q z^q$ which is minimum-phase but *not strictly* minimum-phase. The zeroes of the polynomial matrix $\tilde{\underline{b}}(z) = b_0 + b_1 \rho z + \cdots + (b_q \rho^q) z^q$ are equal to the scaled zeroes of $\underline{b}(z)$, since $\tilde{\underline{b}}(z) = 0$ if and only if $\underline{b}(\rho z) = 0$. Hence, for $\rho < 1$, the matrix $\tilde{\underline{b}}(z)$ satisfies the strict minimum-phase condition. By this trick, we may find in each neighborhood of $\underline{b}$ a strictly minimum-phase matrix $\tilde{\underline{b}}$.

Therefore, the subset

$$\Theta_a \cap \Theta_b \cap \Theta_S \cap \Theta_m \subseteq \tilde{\Theta}_{p,q}$$

of $\tilde{\Theta}_{p,q}$ is open (in $\mathbb{R}^{n^2(p+q)}$ and in $\Theta_{p,q}$) and dense in $\Theta_{p,q}$.   □

Above, we have explained how to construct the ARMA polynomials $\underline{a}(z)$ and $\underline{b}(z)$ from the Smith–McMillan form of a rational filter $\underline{k}(z)$; see also Eq. (6.12). Now, we present an alternative procedure, which starts from the coefficients of the MA(∞) representation of the process. In Exercise 6.2, we have shown how to compute the coefficients of the MA(∞) representations, given the coefficients of the ARMA polynomials $\underline{a}(z)$ and $\underline{b}(z)$. Now, this procedure is reversed. If we set $m = \max(p, q)$ and $a_i = 0$ for $i > p$ and $b_i = 0$ for $i > q$, then we get from Eq. (6.7)

$$(k_{m+1}, k_{m+2}, k_{m+3}, \ldots) = (a_1, a_2, \ldots, a_m) \begin{pmatrix} k_m & k_{m+1} & k_{m+1} & \cdots \\ & & \vdots & \\ k_1 & k_2 & k_2 & \cdots \end{pmatrix}$$

$$= (a_1, a_2, \ldots, a_m) K^m.$$

If the matrix $K^m \in \mathbb{R}^{mn \times \infty}$ has full rank ($= mn$), then the AR coefficients $a_1, \ldots a_m$ are uniquely determined from these equations. Then, for given $\underline{a}(z) = I_n - a_1 z - \cdots a_p z^p$, the coefficients of the MA polynomial are determined from the condition $\underline{b}(z) = \underline{a}(z)\underline{k}(z)$, i.e. by

$$k_0 = I_n = b_0$$
$$k_1 - a_1 k_0 = b_1$$

$$\vdots$$

$$k_m - a_1 k_{m-1} - \cdots - a_m k_0 = b_m.$$

It can be shown that the rank condition is fulfilled on a generic set (i.e. on a set containing an open and dense subset) of ARMA(p, q) systems. Furthermore, one may generalize this procedure, also to the case where this rank condition does not hold. For this generalized procedure, see, for example, Hannan and Deistler (2012, Sect. 2.5). An analogous procedure, for state-space systems, will be discussed in the next chapter; see Proposition 7.11.

References

G.E.P. Box, G.M. Jenkins, *Time Series Analysis: Forecasting and Control*, Revised Edition (Holden-Day, San Francisco, 1976)

E.P. Caines, *Linear Stochastic Systems* (Wiley, New York, 1988)

E.J. Hannan, The identification problem for multiple equation systems with moving average errors. Econometrica **39**(5), 751–765 (1971). ISSN 00129682, 14680262. http://www.jstor.org/stable/1909577

E.J. Hannan, M. Deistler, *The Statistical Theory of Linear Systems*. Classics in Applied Mathematics (SIAM, Philadelphia, 2012). (Originally published: Wiley, New York, 1988)

T. Kailath, *Linear Systems* (Prentice Hall, Englewood Cliffs, New Jersey, 1980)

L. Ljung, *System Identification: Theory for the User*, 2nd edn. (Prentice Hall, 1999)

H. Lütkepohl, *New Introduction to Multiple Time Series Analysis* (Springer, Berlin, 2005)

G.C. Reinsel, *Elements of Multivariate Time Series Analysis* (Springer, 1997)

W. Scherrer, M. Deistler, Chap. 6—vector autoregressive moving average models, in *Conceptual Econometrics Using R* ed. by H.D. Vinod, C. Rao. Handbook of Statistics, vol. 41 (Elsevier, 2019), pp. 145–191. https://doi.org/10.1016/bs.host.2019.01.004. http://www.sciencedirect.com/science/article/pii/S0169716119300045

T. Söderström, P. Stoica,*System Identification* (Prentice Hall, 1989)

State-Space Systems

7

Linear state-space systems, like ARMA systems, are models for stationary processes, more precisely for the class of stationary processes with rational spectral density. ARMA models and state-space models (with white noise as input) represent the same class of stationary processes. State-space systems became particularly popular through the work of Rudolf Kálmán[1] (see e.g. Kalman (1963, 1965, 1974) and Kalman et al. (1969)). They contain an, in general, unobserved variable, the state, which contains all information from the past of the process that is relevant for the future. State-space systems lead to the Kalman filter discussed in this chapter. They are applied much more frequently in control theory than the equivalent ARMA systems.

Two key results in this chapter are the equivalence of controllability and observability with minimality and the description of equivalence classes of observationally equivalent minimal systems. Further, we provide a construction for obtaining a state-space system from the Wold decomposition.

In Sect. 7.4, we will discuss the Kalman filter that goes back to Kalman (1960). The Kalman filter is an algorithm for estimating the unobserved state from observations and for predicting these observations. The Kalman filter is of significant importance for prediction and maximum likelihood estimation.

In terms of general literature for state-space systems, we recommend Kailath (1980), Hannan and Deistler (2012) and Lindquist and Picci (2015). Anderson and Moore (2005) is a classic with regard to the Kalman filter, see also Gómez (2016).

---

[1] Rudolf Kálmán (1930–2016). Born in Hungary, active in the USA and Switzerland. Established modern systems theory. The Kalman filter named after him is one of the most widely used algorithms for prediction and filtering.

© Springer Nature Switzerland AG 2022
M. Deistler and W. Scherrer, *Time Series Models*, Lecture Notes in Statistics 224,
https://doi.org/10.1007/978-3-031-13213-1_7

7.1   Linear State-Space Systems in Innovations Form

In this chapter, with the exception of Sect. 7.4 on the Kalman filter, we consider state-space systems of the form

$$s_{t+1} = As_t + B\epsilon_t \tag{7.1}$$

$$x_t = Cs_t + \epsilon_t, \quad t \in \mathbb{Z} \tag{7.2}$$

where s_t is the m-dimensional state, ϵ_t is the n-dimensional input and x_t is the n-dimensional output. The matrices $A \in \mathbb{R}^{m \times m}$, $B \in \mathbb{R}^{m \times n}$ and $C \in \mathbb{R}^{n \times m}$ are parameters and m is called state dimension. The state s_t is a latent variable, i.e. s_t is unobserved. We also refer to the associated matrix triple (A, B, C) as a state-space system.

We will also assume, unless it is mentioned explicitly, that (ϵ_t) is unobserved white noise with the covariance matrix $\mathbf{E}\epsilon_t \epsilon_t' = \Sigma$. Under the stability condition

$$\varrho(A) < 1 \tag{7.3}$$

$(\varrho(A)$ denotes the spectral radius of A), the unique stationary solution of (7.1) and (7.2) is of the form

$$(s_t) = (I_m L^{-1} - A)^{-1} B(\epsilon_t) \tag{7.4}$$

$$(x_t) = (C(I_m L^{-1} - A)^{-1} B + I_n)(\epsilon_t). \tag{7.5}$$

The transfer function of the filter $(C(I_m L^{-1} - A)^{-1} B + I_n)$ is

$$\underline{k}(z) = C(I_m z^{-1} - A)^{-1} B + I_n \tag{7.6}$$

and the impulse response, i.e. the coefficients of the power series expansion $\underline{k}(z) = \sum_{j \geq 0} k_j z^j$, are

$$k_0 = I_n, \quad k_j = CA^{j-1}B \text{ for } j > 0. \tag{7.7}$$

Clearly, the transfer function is rational and stable (i.e. it has no poles for $|z| \leq 1$); hence, the process (x_t) has a rational spectral density.

With the help of the Woodbury matrix identity, one obtains the following formula for the inverse transfer function $\underline{k}^{-1}(z)$:

$$\underline{k}^{-1}(z) = \left(I + C(z^{-1}I_m - A)^{-1}B\right)^{-1} = I - C(z^{-1}I_m - A + BC)^{-1}B.$$

That is, $\underline{k}^{-1}$ is the transfer function of the state-space system $(A - BC, B, -C)$. We can also derive this representation directly from Eqs. (7.1) and (7.2) by first expressing ϵ_t by s_t and x_t and then inserting it into the state Eq. (7.1):

$$\epsilon_t = x_t - Cs_t \tag{7.8}$$

$$s_{t+1} = As_t + B\epsilon_t = (A - BC)s_t + Bx_t. \tag{7.9}$$

If, in addition, the minimum-phase condition

$$\varrho(A - BC) \leq 1 \tag{7.10}$$

holds, then the transfer function $\underline{k}$ is also minimum-phase (i.e. the transfer function has no zeros for $|z| < 1$). The transfer function $\underline{k}$ then corresponds to the Wold representation of the process (see Lemma 6.1) and the ϵ_t's are the innovations. This motivates the following definition:

Definition 7.1 A state-space system (A, B, C) that satisfies the stability condition (7.3) and the minimum-phase condition (7.10) is called a *state-space system in innovations form.*

We will also often impose the slightly stronger strict minimum-phase condition

$$\varrho(A - BC) < 1. \tag{7.11}$$

Prediction is particularly simple for a system in innovations form. From Eqs. (7.1), (7.2), it follows immediately that

$$s_{t+1} = \sum_{j \geq 0} A^j B \epsilon_{t-j}$$

$$x_{t+h} = C A^{h-1} s_{t+1} + \epsilon_{t+h} + C B \epsilon_{t+h-1} + \cdots + C A^{h-2} B \epsilon_{t+1}.$$

Thus, the h-step ahead prediction from the infinite past is

$$\hat{x}_{t,h} = C A^{h-1} s_{t+1},$$

since

$$s_{t+1} = \sum_{j \geq 0} A^j B \epsilon_{t-j} \in (\mathbb{H}_t(\epsilon))^m = (\mathbb{H}_t(x))^m$$

and $(x_{t+h} - C A^{h-1} s_{t+1}) \perp \mathbb{H}_t(\epsilon) = \mathbb{H}_t(x)$. Hence, the *(finite-dimensional)* state s_{t+1} contains all the information from the past that is relevant for the future.

7.2   Controllability, Observability and Minimality of State-Space Systems

We now consider a state-space system (A, B, C), initially making no further assumptions such as the stability condition and also no assumptions about the input process. In this case, a stationary solution does not necessarily exist. Therefore, we consider solutions on $\mathbb{N}$ with zero initial state. By recursively solving Eqs. (7.1), (7.2) for $t = 1, 2, \ldots$ it follows immediately that

$$s_t = As_{t-1} + B\epsilon_{t-1} = A^t s_0 + A^{t-1}B\epsilon_0 + A^{t-2}B\epsilon_1 + \cdots + B\epsilon_{t-1}$$

$$x_t = Cs_t + \epsilon_t = CA^t s_0 + \epsilon_t + CB\epsilon_{t-1} + CAB\epsilon_{t-2} + \cdots + CA^{t-1}B\epsilon_0$$

$$= CA^t s_0 + \sum_{j=0}^{t} k_j \epsilon_{t-j}.$$

If we start the system with the initial state $s_0 = 0$, then the outputs x_t, $t > 0$ are determined by the inputs ϵ_t, $t \geq 0$ and the transfer function $\underline{k}(z)$ (see (7.6)) resp. the impulse response $(k_j \mid j \geq 0)$ (see (7.7)).

Clearly, the transfer function is determined by the state-space system (A, B, C); however, the converse is not true (in general). Even the state-space dimension m is not uniquely determined by $\underline{k}(z)$. In the following, we analyze the relation between the transfer function (resp. impulse response) and state-space system in more detail.

Definition 7.2 A state-space system (A, B, C) is called *controllable*, if

$$\text{rk } \underbrace{(B, AB, \ldots, A^{m-1}B)}_{\mathcal{C} \in \mathbb{R}^{m \times mn}} = m, \tag{7.12}$$

and it is called *observable*, if

$$\text{rk } \underbrace{(C', A'C', \ldots, (A')^{m-1}C')'}_{\mathcal{O} \in \mathbb{R}^{mn \times m}} = m. \tag{7.13}$$

A state-space model is called *minimal*, if its state-space dimension m is minimal among all state-space systems with the same transfer function.

One can interpret controllability and observability for the case of non-stochastic, controlled and observed inputs, respectively, as follows. In analogy to the above, the following holds for $t \geq 0$:

$$s_{t+m} = A^m s_t + \underbrace{(B, AB, \ldots, A^{m-1}B)}_{\mathcal{C}}(\epsilon'_{t+m-1}, \epsilon'_{t+m-2}, \ldots, \epsilon'_t)'.$$

The *controllability* of the system (i.e. rk $\mathcal{C} = m$) thus implies that the system can be moved—by suitable choice of inputs – from a state s_t in m time steps to any state $s^* = s_{t+m}$.

One can also easily show that

$$
\begin{pmatrix} x_t \\ x_{t+1} \\ \vdots \\ x_{t+m-1} \end{pmatrix} = \underbrace{\begin{pmatrix} C \\ CA \\ \vdots \\ CA^{m-1} \end{pmatrix}}_{\mathcal{O}} s_t + \begin{pmatrix} k_0 & 0 & \cdots & 0 \\ k_1 & k_0 & \cdots & 0 \\ \vdots & \vdots & \ddots & \vdots \\ k_{m-1} & k_{m-2} & \cdots & k_0 \end{pmatrix} \begin{pmatrix} \epsilon_t \\ \epsilon_{t+1} \\ \vdots \\ \epsilon_{t+m-1} \end{pmatrix}.
$$

Therefore, one can determine the state s_t for an *observable* system ($\mathrm{rk}(\mathcal{O}) = m$) if the future outputs x_{t+j} and inputs ϵ_{t+j} for $j = 0, \ldots, m-1$ are known.

For the following analyses, we define

$$
\mathcal{C}_\infty : = (B, AB, A^2 B \ldots) \in \mathbb{R}^{m \times \infty}
$$

$$
\mathcal{O}_\infty : = (C', A'C', (A')^2 C' \ldots)' \in \mathbb{R}^{\infty \times m}
$$

$$
\mathcal{H}_\infty : = \mathcal{O}_\infty \mathcal{C}_\infty = \begin{pmatrix} k_1 & k_2 & k_3 & \cdots \\ k_2 & k_3 & k_4 & \cdots \\ k_3 & k_4 & k_5 & \cdots \\ \vdots & \vdots & \vdots & \end{pmatrix} \in \mathbb{R}^{\infty \times \infty}
$$

$$
\mathcal{H}_m : = \mathcal{O}\mathcal{C} \in \mathbb{R}^{nm \times nm}.
$$

The matrix $\mathcal{H}_\infty$ is the so-called *Hankel matrix of the transfer function* $\underline{k}(z)$ and $\mathcal{H}_m$ is the left upper ($mn \times mn$) dimensional submatrix of $\mathcal{H}_\infty$. According to the Cayley–Hamilton theorem, coefficients $d_j^{(k)} \in \mathbb{R}$ exist such that for $k \geq 0$

$$
A^k = d_0^{(k)} I_m + d_1^{(k)} A + \cdots + d_{m-1}^{(k)} A^{m-1}.
$$

This allows us to construct a matrix $D \in \mathbb{R}^{nm \times \infty}$ in such a way that $\mathcal{C}_\infty = \mathcal{C}D$, $\mathcal{O}_\infty = D'\mathcal{O}$ and $\mathcal{H}_\infty = D'\mathcal{H}_m D$. This implies

$$
\mathrm{col}(\mathcal{C}_\infty) = \mathrm{col}(\mathcal{C}), \ \ \mathrm{row}(\mathcal{O}_\infty) = \mathrm{row}(\mathcal{O}) \ \ \text{und} \ \ \mathrm{rk}(\mathcal{H}_\infty) = \mathrm{rk}(\mathcal{H}_m) \leq m,
$$

where $\mathrm{col}(M)$ ($\mathrm{row}(M)$) denotes the column space (resp. row space) of a matrix M.

Proposition 7.3 *The following statements are equivalent for a state-space system* (A, B, C) *with state-space dimension m:*

(1) The system is observable and controllable.
(2) The system is minimal.
(3) $\mathrm{rk}(\mathcal{H}_m) = m$.
(4) $\mathrm{rk}(\mathcal{H}_\infty) = m$.

Proof (1) $\Rightarrow$ (3): It follows from the observability and controllability that $\mathcal{O}'\mathcal{O}$ and $\mathcal{C}\mathcal{C}'$ are non-singular $m \times m$ matrices. Thus, this also holds for $\mathcal{O}'\mathcal{O}\mathcal{C}\mathcal{C}'$ and $\mathcal{H}_m = \mathcal{O}\mathcal{C}$ must therefore have rank m.

(3) $\Rightarrow$ (2): Suppose (A, B, C) is not minimal, then there exists a state-space system $(\bar{A}, \bar{B}, \bar{C})$ with the same transfer function and with $\bar{A} \in \mathbb{R}^{\bar{m} \times \bar{m}}$ with $\bar{m} < m$. However, then it follows that $\mathrm{rk}(\mathcal{H}_m) \leq \bar{m} < m$.

(2) $\Rightarrow$ (1): Suppose, for example, that the system is not controllable. As can be directly seen from (7.6) (resp. (7.7)), the transfer function does not change if the following parameter transformation is performed

$$\bar{A} = TAT^{-1} \tag{7.14}$$
$$\bar{B} = TB \tag{7.15}$$
$$\bar{C} = CT^{-1} \tag{7.16}$$

with $T \in \mathbb{R}^{m \times m}$, $\det T \neq 0$. Now, if $\mathrm{rk}(\mathcal{C}) = \mathrm{rk}(\mathcal{C}_\infty) = \bar{m} < m$, then there exists a non-singular matrix T, such that the last $m - \bar{m}$ rows of $T\mathcal{C}_\infty = \bar{\mathcal{C}}_\infty = (\bar{B}, \bar{A}\bar{B}, \bar{A}^2 \bar{B}, \ldots)$ are equal to zero. Now we partition $\bar{A}$, $\bar{B}$, $\bar{C}$ and $\bar{\mathcal{C}}_\infty$ into blocks with $\bar{m}$ and $m - \bar{m}$ rows resp. columns:

$$\bar{A} = \begin{pmatrix} \bar{A}_{11} & \bar{A}_{12} \\ \bar{A}_{21} & \bar{A}_{22} \end{pmatrix}, \ \bar{B} = \begin{pmatrix} \bar{B}_1 \\ \bar{B}_2 \end{pmatrix}, \ \bar{C} = (\bar{C}_1, \bar{C}_2), \ \bar{\mathcal{C}}_\infty = \begin{pmatrix} \bar{\mathcal{C}}_{1,\infty} \\ \bar{\mathcal{C}}_{2,\infty} \end{pmatrix}.$$

The relation $\bar{\mathcal{C}}_\infty = (\bar{B}, \bar{A}\bar{\mathcal{C}}_\infty)$ then implies together with $\mathrm{rk}(\bar{\mathcal{C}}_{1,\infty}) = \bar{m}$ and $\bar{\mathcal{C}}_{2,\infty} = 0$ that $\bar{B}_2 = 0 \in \mathbb{R}^{m-\bar{m} \times n}$ and $\bar{A}_{21} = 0 \in \mathbb{R}^{m-\bar{m} \times \bar{m}}$. From this, it now follows that

$$k_1 = CB = \bar{C}_1\bar{B}_1, \ k_2 = CAB = \bar{C}_1\bar{A}_{11}\bar{B}_1, \ k_3 = CA^2B = \bar{C}_1\bar{A}_{11}^2\bar{B}_1, \ldots$$

Hence, the system $(\bar{A}_{11}, \bar{B}_1, \bar{C}_1)$ has the same transfer function as (A, B, C), but a smaller state-space dimension.

If the system is not observable, one can argue in a completely analogous way.

(3) $\Leftrightarrow$ (4): follows directly from $\mathrm{rk}(\mathcal{H}_m) = \mathrm{rk}(\mathcal{H}_\infty)$. $\quad\square$

The proof shows how an arbitrary state-space system can be transformed into a minimal system. We also see that the minimal system constructed in this way satisfies the stability condition (the (strict) minimum-phase condition) if the original (non-minimal) system satisfies the stability condition (or the (strict) minimum-phase condition).

Proposition 7.4 *Two minimal state-space systems $(\bar{A}, \bar{B}, \bar{C})$ and (A, B, C) have the same transfer function (i.e., they are observationally equivalent) if and only if there exists a non-singular matrix T, such that (7.14)–(7.16) holds.*

Proof One direction is evident. Conversely, for two minimal systems that have the same transfer function, it follows (in evident notation) that

$$\mathcal{O}\mathcal{C} = \bar{\mathcal{O}}\bar{\mathcal{C}}$$

and thus

$$\bar{\mathcal{C}} = (\bar{\mathcal{O}}'\bar{\mathcal{O}})^{-1}\bar{\mathcal{O}}'\mathcal{O}\mathcal{C} \tag{7.17}$$

and

$$\bar{\mathcal{O}} = \mathcal{O}\mathcal{C}\bar{\mathcal{C}}'(\bar{\mathcal{C}}\bar{\mathcal{C}}')^{-1}. \tag{7.18}$$

If we now write $T = (\bar{\mathcal{O}}'\bar{\mathcal{O}})^{-1}\bar{\mathcal{O}}'\mathcal{O}$ and $S = \mathcal{C}\bar{\mathcal{C}}'(\bar{\mathcal{C}}\bar{\mathcal{C}}')^{-1}$, it follows that $\mathcal{O}\mathcal{C} = \bar{\mathcal{O}}\bar{\mathcal{C}} = \mathcal{O}STC$. Furthermore, $I_n = ST$ must hold due to observability and controllability, i.e. T is non-singular and $S = T^{-1}$. Equations (7.15) and (7.16) follow immediately from (7.17) and (7.18). Finally,

$$\bar{\mathcal{O}}\bar{A}\bar{\mathcal{C}} = \begin{pmatrix} k_2 & k_3 & \cdots \\ k_3 & k_4 & \cdots \\ \vdots & \vdots & \end{pmatrix} = \mathcal{O}A\mathcal{C}$$

implies the transformation (7.14).   □

As can be easily seen, a transformation T in (7.14)–(7.16) corresponds to a transformation

$$\bar{s}_t = T s_t \tag{7.19}$$

of the (minimal) states.

At the end of this section, we shall discuss the relationship of the poles (zeros) of the transfer function to the eigenvalues of matrix A (resp. matrix $(A - BC)$). For this purpose, we first need an alternative characterization of observability or controllability.

Lemma 7.5 *A state-space system* (A, B, C) *is controllable if and only if* $((\lambda I - A), B)$ *(as polynomials in* $(\lambda \in \mathbb{C}))$ *is relatively left prime and it is observable if and only if* $((\lambda I - A)', C')$ *is relatively left prime.*

Proof The pair $((\lambda I - A), B)$ is not left prime if and only if there exists a $\lambda_0 \in \mathbb{C}$ and a vector $u \in \mathbb{C}^{1 \times m}$, $u \neq 0$ such that $u(\lambda_0 I_m - A, B) = 0$. That is, u is a left eigenvector of A which is also an element of the left kernel of B. Clearly, then it holds that $u\mathcal{C} = (uB, uAB, \ldots, uA^{m-1}B) = 0$, that is, (A, B, C) is not controllable. Conversely, if (A, B, C) is not controllable, then $0 = u\mathcal{C} = (uB, uAB, \ldots, uA^{m-1}B)$ and therefore $vB = 0$ for all elements of the Krylov subspace $\mathrm{sp}\{uA^j \mid j \geq 0\}$. The Krylov subspace contains at least one left eigenvector of A and thus, the pair $((\lambda I - A), B)$ cannot be relatively left prime.

The argument for "observable" is quite analogous.   □

Lemma 7.6 *For minimal systems* (A, B, C), *it holds that* z_0 *is a pole of the transfer function* $\underline{k}(z) = I_n + C(I_m z^{-1} - A)^{-1} B$ *if and only if* $\lambda_0 = z_0^{-1}$ *is a nonzero eigenvalue of* A.

Proof If $\lambda_0 = z_0^{-1}$ is not an eigenvalue of A, then $(I_n z_0^{-1} - A)$ is invertible and z_0 is therefore not a pole of $\underline{k}(z)$.

The pair $((\lambda I - A), B)$ is left prime, and hence there exists a polynomial right inverse (see e.g. Hannan and Deistler (2012, Lemma (2.2.1))), i.e. polynomial matrices $g(\lambda)$ and $h(\lambda)$ such that $(\lambda I_m - A)g(\lambda) + Bh(\lambda) = I_m$. From this, we obtain

$$(\lambda I_m - A)^{-1} = g(\lambda) + (\lambda I_m - A)^{-1} Bh(\lambda).$$

The polynomial matrices g and h are finite for all $\lambda \in \mathbb{C}$ and therefore each eigenvalue λ_0 of A is a pole of $(\lambda I_m - A)^{-1} B$. Since $((\lambda I - A)', C')$ is left prime, $\tilde{g}(\lambda)(I_m \lambda - A) + \tilde{h}(\lambda)C = I_m$ follows analogously for two suitable polynomial matrices $\tilde{g}$ and $\tilde{h}$. Thus,

$$(\lambda I_m - A)^{-1} B = \tilde{g}(\lambda)B + \tilde{h}(\lambda)C(\lambda I_m - A)^{-1} B$$

and we see that each eigenvalue λ_0 of A must also be a pole of $C(\lambda I_m - A)^{-1} B$.

Furthermore, we note that $\underline{k}(0) = I_n$ holds and that $z_0 = 0$ is hence not a pole of $\underline{k}(z)$. □

The zeros of the transfer function $\underline{k}$ are the poles of the inverse transfer function

$$\underline{k}^{-1}(z) = \left(I + C(z^{-1} I_m - A)^{-1} B \right)^{-1} = I - C(z^{-1} I_m - A + BC)^{-1} B.$$

$\underline{k}^{-1}(z)$ is, as already explained above, the transfer function of the state-space system $(A - BC, B, -C)$.

Lemma 7.7 *A state-space system* (A, B, C) *is minimal if and only if the system* $(A - BC, B, -C)$ *is minimal.*

Proof If (A, B, C) is a non-minimal state-space system with state dimension m, then there exists a system $(\bar{A}, \bar{B}, \bar{C})$ with a smaller state-space dimension $\bar{m} < m$ that describes the same transfer function $\underline{k}$. Hence, $(A - BC, B, -C)$ and $(\bar{A} - \bar{B}\bar{C}, \bar{B}, -\bar{C})$ are two systems that have the same transfer function $\underline{k}^{-1}$ and $(A - BC, B, -C)$ is thus non-minimal. Analogously, the non-minimality of (A, B, C) follows from the non-minimality of $(A - BC, B, -C)$. □

Exercise 7.8 Show that a system (A, B, C) is observable (controllable) if and only if the system $(A - BC, B, -C)$ is observable (controllable). Note:

$$[\lambda I - A + BC, B] = [\lambda I - A, B] \begin{pmatrix} I & 0 \\ C & I \end{pmatrix}.$$

This observation can be used to prove the above lemma.

The above lemma together with Lemma 7.6 thus yields the following characterization of the zeros of the transfer function $\underline{k}$.

Lemma 7.9 *For minimal systems* (A, B, C), *it holds that* z_0 *is a zero of the transfer function* $\underline{k}(z) = I_n + C(I_m z^{-1} - A)^{-1}B$ *if and only if* $\lambda_0 = z_0^{-1}$ *is a nonzero eigenvalue of* $(A - BC)$.

We conclude this section by returning to the case where the input (ϵ_t) is unobserved white noise. We have shown: If the system (A, B, C) is minimal, then this system corresponds to the Wold representation of the process if and only if the system is in innovations form, i.e. the stability condition (7.3) and the minimum-phase condition (7.10) are satisfied. If the system is not minimal, then these conditions are sufficient but not necessary.

Now suppose that the system is in innovations form (but not necessarily minimal). We consider the predictions for the future random variables x_{t+h}, $h > 0$ for given, present and past variables x_r, $r \le t$. That is, we consider the projections of x_{t+h}, $h > 0$ onto the "present and past of the process", i.e. onto the space $\mathbb{H}_t(x)$. Using our notation, it follows that

$$
\begin{pmatrix} \hat{x}_{t,1} \\ \hat{x}_{t,2} \\ \hat{x}_{t,3} \\ \vdots \end{pmatrix} = \begin{pmatrix} C \\ CA \\ CA^2 \\ \vdots \end{pmatrix} s_{t+1} = \mathcal{O}_\infty \mathcal{C}_\infty \underbrace{\begin{pmatrix} \epsilon_t \\ \epsilon_{t-1} \\ \epsilon_{t-2} \\ \vdots \end{pmatrix}}_{:=\epsilon_t^\infty} = \mathcal{H}_\infty \epsilon_t^\infty.
$$

The space $\mathbb{H}_t^+(x) := \overline{\mathrm{sp}}\{\hat{x}_{t,h} \mid h > 0\}$ which is spanned by these predictions is therefore a subspace of the space spanned by the components of the state s_{t+1} and is hence finite dimensional. In the next section, we will also show the reverse direction. If the so-called *predictor space* $\mathbb{H}_t^+(x) := \overline{\mathrm{sp}}\{\hat{x}_{t,h} \mid h > 0\}$ is finite dimensional, then the process has a state-space representation.

If the covariance matrix $\Sigma = \mathbf{E}\epsilon_t \epsilon_t' > 0$ is positive definite, then a state-space system in innovations form is minimal if and only if the components of s_{t+1} form a basis for the space $\mathbb{H}_t^+(x) := \overline{\mathrm{sp}}\{\hat{x}_{t,h} \mid h > 0\}$. Under the assumption that $\Sigma > 0$, the covariance matrix $\mathbf{E}s_{t+1}s_{t+1}' = \mathcal{C}_\infty \mathrm{diag}(\Sigma, \Sigma, \ldots)\mathcal{C}_\infty'$ of the state has the same rank as $\mathcal{C}_\infty$. Here $\mathrm{diag}(\Sigma, \Sigma, \ldots)$ denotes an infinite-dimensional block diagonal matrix with diagonal blocks equal to Σ. Hence, the components $s_{k,t+1}$ are linearly independent if and only if the system is controllable. If the system is observable, then it follows that $s_{t+1} = (\mathcal{O}'\mathcal{O})^{-1}\mathcal{O}'(\hat{x}_{t,1}', \ldots, \hat{x}_{t,m}')'$ and therefore $\mathbb{H}_t^+(x) = \mathrm{sp}\{s_{t+1}\}$. However, if the system is not observable, then the dimension of $\mathbb{H}_t^+(x)$ is less than m and hence either $\mathbb{H}_t^+(x)$ is a proper subspace of $\mathrm{sp}\{s_{t+1}\}$ or the components $s_{k,t+1}$ are not linearly independent.

7.3   From the Wold Representation to a State-Space System

First we consider general rational transfer functions $\underline{k}(z)$, for which $\underline{k}(0) = I_n$, i.e. we do not impose the stability or the minimum-phase condition. We want to show that such rational transfer functions can be "realized" by a state-space system (A, B, C), i.e. there exist matrices A, B, C such that $\underline{k}(z) = (C(I_m z^{-1} - A)^{-1} B + I_n)$.

Proposition 7.10 *For any rational transfer function $\underline{k}(z)$ (with $\underline{k}(0) = I_n$), the rank of the corresponding Hankel matrix $\mathcal{H}_\infty$ is finite.*

Proof We write

$$\underline{k}(z) = \underline{a}^{-1}(z)\underline{b}(z)$$

where $\underline{a}(z) = \sum_{j=0}^{p} a_j z^j$ and $\underline{b}(z) = \sum_{j=0}^{p} b_j z^j$ are polynomial matrices of maximum degree p. The polynomials $\underline{a}, \underline{b}$ can be determined e.g. from the Smith–McMillan form (4.22) of the transfer function $\underline{k}$. We assume without loss of generality that $a_0 = \underline{a}(0) = \underline{b}(0) = b_0 = I_n$. From

$$(I_n + b_1 z + \cdots + b_p z^p) = \underline{b}(z) = \underline{a}(z)\underline{k}(z) = (I_n + a_1 z + \cdots + a_p z^p)(I_n + k_1 z + k_2 z^2 + \cdots)$$

it follows that

$$(a_p, a_{p-1}, \ldots, I_n) \begin{pmatrix} k_1 & k_2 & \cdots \\ k_2 & k_3 & \cdots \\ \vdots & \vdots & \\ k_{p+1} & k_{p+2} & \cdots \end{pmatrix} = (0, 0, \ldots)$$

and hence the assertion follows from the (block Hankel) structure of $\mathcal{H}_\infty$.   □

In the following, we will describe a construction to obtain a state-space system (A, B, C) from a Hankel matrix $\mathcal{H}_\infty$ with rank m. (This procedure is the "state-space" analogon to the construction of an ARMA system, which has been given at the end of Sect. 6.3.) Let $S \in \mathbb{R}^{m \times \infty}$ be a matrix, such that the rows of $S\mathcal{H}_\infty$ form a basis for the row space of $\mathcal{H}_\infty$. We now determine (A, B, C) from

$$S \begin{pmatrix} k_2 & k_3 & \cdots \\ k_3 & k_4 & \cdots \\ \vdots & \vdots & \end{pmatrix} = AS\mathcal{H}_\infty \tag{7.20}$$

$$B = S \begin{pmatrix} k_1 \\ k_2 \\ \vdots \end{pmatrix} \tag{7.21}$$

$$(k_1, k_2, \ldots) = CS\mathcal{H}_\infty. \tag{7.22}$$

From these equations, $k_1 = CB$ follows as desired and for $j > 1$

$$k_j = CS \begin{pmatrix} k_j \\ k_{j+1} \\ \vdots \end{pmatrix} = CAS \begin{pmatrix} k_{j-1} \\ k_j \\ \vdots \end{pmatrix} = \cdots = CA^{j-1}S \begin{pmatrix} k_1 \\ k_2 \\ \vdots \end{pmatrix} = CA^{j-1}B.$$

The system (A, B, C) is minimal due to Proposition 7.3. We have thus proved the following proposition:

Proposition 7.11 *For every transfer function* $\underline{k}(z) = \sum_{j \geq 0} k_j z^j$, $\underline{k}(0) = I_n$, *where the corresponding Hankel matrix has finite rank* m *there exists a (minimal) state-space system* (A, B, C) *with state dimension* m, *such that* $\underline{k}(z) = C(I_m z^{-1} - A)^{-1}B + I_n$. *Therefore the transfer function is rational. Note that* $\underline{k}(z)$ *here is assumed to have no poles at* $z = 0$.

The construction given above for (A, B, C) is unique for given S. It is easy to see that S is unique except for pre-multiplication with non-singular matrices T. This corresponds to the state transformation (7.19). Identifiability for (A, B, C) is obtained by choosing a unique matrix S. A commonly used choice is the selection matrix corresponding to the first m linearly independent rows of $(\mathcal{H}_\infty)$, see Hannan and Deistler (2012). We will not discuss issues related to identifiability of the system parameters (A, B, C) here.

Now let

$$x_t = \sum_{j=0}^\infty k_j \epsilon_{t-j} \tag{7.23}$$

be the Wold representation of the process (x_t) and

$$\underline{k}(z) = \sum_{j=0}^\infty k_j z^j, \quad k_0 = I_n \tag{7.24}$$

be the corresponding transfer function.

Lemma 7.12 Given the Wold representation (7.23) of a regular process the following statements are equivalent:

(1) The transfer function $\underline{k}(z) = \sum_{j \geq 0} k_j z^j$ is rational.
(2) The Hankel matrix $\mathcal{H}_\infty = (k_{i+j-1})_{i,j \geq 1}$ has finite rank.
(3) The predictor space $\mathbb{H}_t^+(x) = \overline{\mathrm{sp}}\{\hat{x}_{t,h} \mid h > 0\}$ is finite dimensional.

Proof Due to Propositions 7.10 and 7.11 we only have to show that points (2) and (3) are equivalent. However, this follows directly from

$$
\begin{pmatrix} \hat{x}_{t,1} \\ \hat{x}_{t,2} \\ \hat{x}_{t,3} \\ \vdots \end{pmatrix} = \mathcal{H}_\infty \begin{pmatrix} \epsilon_t \\ \epsilon_{t-1} \\ \epsilon_{t-2} \\ \vdots \end{pmatrix}.
$$

$\square$

If the transfer function $\underline{k}$ is rational, then $\underline{k}$ is stable and minimum-phase (see Corollary 6.8) and, according to Lemmas 7.6 and 7.9, the associated (minimal) state-space system (A, B, C) is therefore in innovations form.

There is a Hilbert space construction analogous to the construction above, see Akaike (1974): Suppose the predictor space $\mathbb{H}_t^+(x)$ is m-dimensional. Now we choose an m-dimensional random vector $s_{t+1} = S(\hat{x}_{t,1}, \hat{x}_{t,2}, \ldots, \hat{x}_{t,o})$, $S \in \mathbb{R}^{m \times no}$, whose components $s_{k,t+1}$ form a basis for $\mathbb{H}_t^+(x)$. Clearly, $(s_t = S(\hat{x}'_{t-1,1}, \ldots, \hat{x}'_{t-1,o})$ $\mid t \in \mathbb{Z})$ is a stationary process and $\{s_{1r}, \ldots, s_{mr}\}$ is a basis for $\mathbb{H}_{r-1}^+(x)$ for all $r \in \mathbb{Z}$. By construction, it holds that $\hat{x}_{t-1,1} \in (\mathbb{H}_{t-1}^+(x))^n$ and hence there exists a matrix $C \in \mathbb{R}^{n \times m}$ such that

$$
x_t = \hat{x}_{t-1,1} + \epsilon_t = C s_t + \epsilon_t.
$$

The space $\mathbb{H}_t(x)$ is the sum of the mutually orthogonal spaces $\mathbb{H}_{t-1}(x)$ and $\mathrm{sp}\{\epsilon_t\}$. Therefore, it now follows that

$$
s_{t+1} = S \, \mathrm{P}_{\mathbb{H}_t(x)} \begin{pmatrix} x_{t+1} \\ x_{t+2} \\ \vdots \\ x_{t+o} \end{pmatrix} = S \, \mathrm{P}_{\mathbb{H}_{t-1}(x)} \begin{pmatrix} x_{t+1} \\ x_{t+2} \\ \vdots \\ x_{t+o} \end{pmatrix} + S \, \mathrm{P}_{\mathrm{sp}\{\epsilon_t\}} \begin{pmatrix} x_{t+1} \\ x_{t+2} \\ \vdots \\ x_{t+o} \end{pmatrix}
$$

$$
= A s_t + B \epsilon_t
$$

for suitable matrices $A \in \mathbb{R}^{m \times m}$ and $B \in \mathbb{R}^{m \times n}$.

This construction shows that the minimal state s_t (more precisely, the components of s_t) is a basis for $\mathbb{H}_x^+(t)$.

Proposition 7.13 *The following statements are equivalent:*

(1) (x_t) is a stationary process with a rational spectral density which is of full rank μ-a.e.

(2) (x_t) is a regular ARMA process and the innovation variance is positive definite.

(3) (x_t) is the regular and stationary solution of a state-space system in innovations form and the innovation variance is positive definite.

Proof The transfer function and hence the spectral densities are rational for regular ARMA processes as well as for the stationary solutions of state-space systems in innovations form. Conversely, it follows from Proposition 6.7 that a stable and minimum-phase transfer function corresponds to every rational density. One can then construct an ARMA system (see Proposition 6.13) or a state-space system in innovations form (see proposition 7.11) for this transfer function.   □

Exercise 7.14 Prove: A regular process is an ARMA process if and only if the Hankel matrix of the covariance function

$$\begin{pmatrix} \gamma(1) & \gamma(2) & \gamma(3) & \cdots \\ \gamma(2) & \gamma(3) & \gamma(4) & \cdots \\ \gamma(3) & \gamma(4) & \gamma(5) & \cdots \\ \vdots & \vdots & & \end{pmatrix}$$

has finite rank.

Hint(s): It suffices to prove the claim for the case that the innovation variance $\Sigma = \mathbf{E}\epsilon_t\epsilon_t'$ is positive definite. In this case, the Hankel matrix of the transfer function and the Hankel matrix of the covariance function have the same rank. This may be easily seen from the relations below

$$\begin{pmatrix} x_t \\ x_{t+1} \\ x_{t+2} \\ \vdots \end{pmatrix} = \begin{pmatrix} k_1 & k_2 & k_3 & \cdots \\ k_2 & k_3 & k_4 & \cdots \\ k_3 & k_4 & k_5 & \cdots \\ \vdots & \vdots & \vdots & \end{pmatrix} \begin{pmatrix} \epsilon_{t-1} \\ \epsilon_{t-2} \\ \epsilon_{t-3} \\ \vdots \end{pmatrix} + \begin{pmatrix} I_n & 0 & 0 & \cdots \\ k_1 & I_n & 0 & \cdots \\ k_2 & k_1 & I_n & \\ \vdots & & & \ddots \end{pmatrix} \begin{pmatrix} \epsilon_t \\ \epsilon_{t+1} \\ \epsilon_{t+2} \\ \vdots \end{pmatrix}$$

$$\begin{pmatrix} x_{t-1} \\ x_{t-2} \\ x_{t-3} \\ \vdots \end{pmatrix} = \begin{pmatrix} I_n & k_1 & k_2 & \cdots \\ 0 & I_n & k_1 & \cdots \\ 0 & 0 & I_n & \\ & & & \ddots \end{pmatrix} \begin{pmatrix} \epsilon_{t-1} \\ \epsilon_{t-2} \\ \epsilon_{t-3} \\ \vdots \end{pmatrix}$$

Exercise 7.15 Suppose an ARMA(p, q) process $\underline{a}(L)x_t = \underline{b}(L)\epsilon_t$, where the stability condition and the minimum-phase condition are satisfied. Show that

$$m \leq \max(p, q)n$$

holds for the state-space dimension m of an equivalent minimal state-space system (in innovations form). Note: See the proof of Proposition 7.10.

Exercise 7.16 (*continuation*) Without loss of generality, we assume that $p = q$. Show the following: By defining

$$s_t = (\hat{x}_{t-1,1}', \ldots, \hat{x}_{t-1,p}')'$$

as state, one obtains the following state-space representation (A, B, C) for (x_t):

$$
A = \begin{pmatrix} 0 & I_n & \cdots & 0 \\ \vdots & \vdots & \ddots & \vdots \\ 0 & 0 & \cdots & I_n \\ a_p & a_{p-1} & \cdots & a_1 \end{pmatrix}, \quad B = \begin{pmatrix} k_1 \\ k_2 \\ \vdots \\ k_p \end{pmatrix}
$$
$$
C = \begin{pmatrix} I_n & 0 & \cdots & 0 \end{pmatrix}
$$

where $\underline{k}(z) = \underline{a}^{-1}(z)\underline{b}(z) = \sum_{j\geq 0} k_j z^j$, i.e. $x_t = \sum_{j\geq 0} k_j \epsilon_{t-j}$ is the Wold representation of the process.

Note: Use both the representation $\hat{x}_{t,h} = \sum_{j\geq h} k_j \epsilon_{t+h-j}$ and the representation of the h-step ahead prediction derived in the Exercise 6.5 (prediction of ARMA processes), showing in particular that

$$
\hat{x}_{t-1,p+1} = a_1 \hat{x}_{t-1,p} + \cdots + a_p \hat{x}_{t-1,1}.
$$

This state-space representation for the ARMA process (x_t) also provides a possibility to determine the covariance function. See Exercises 7.28 and 7.29 below.

Exercise 7.17 Suppose a minimal state-space system in innovations form with state dimension m: Show

$$
s_t \in (\text{sp}\{x_{t-1}, \ldots, x_{t-m}, \epsilon_{t-1}, \ldots, \epsilon_{t-m}\})^m.
$$

It is therefore possible to construct an equivalent ARMA (p, q) system with $p, q \leq m$ by projecting x_t onto $\text{sp}\{x_{t-1}, \ldots, x_{t-m}, \epsilon_{t-1}, \ldots, \epsilon_{t-m}\}$.
Note: Verify and apply the following equations (see (7.9)):

$$
s_t = (A - BC)^m s_{t-m} + (B, (A - BC)B, \ldots, (A - BC)^{m-1}B)x_{t-1}^m \qquad (7.25)
$$

$$
\begin{pmatrix} x_{t-m} \\ x_{t-m+1} \\ \vdots \\ x_{t-1} \end{pmatrix} = \mathcal{O}s_{t-m} + \begin{pmatrix} I_n & 0 & \cdots & 0 \\ CB & I_n & & \vdots \\ \vdots & & \ddots & \vdots \\ CA^{m-2}B & \cdots & & I_n \end{pmatrix} \begin{pmatrix} \epsilon_{t-m} \\ \epsilon_{t-m+1} \\ \vdots \\ \epsilon_{t-1} \end{pmatrix}. \qquad (7.26)
$$

Exercise 7.18 Let $(x_t = a_1 x_{t-1} + \cdots + a_p x_{t-p} + \epsilon_t)$ be a regular AR(p) process, where the polynomial matrix $\underline{a}(z) = I_n - a_1 z - \cdots - a_p z^p$ satisfies the stability condition. Verify the following state-space representation for the process (x_t):

$$x_t^p = \begin{pmatrix} a_1 & \cdots & a_{p-1} & a_p \\ I_n & \cdots & 0 & 0 \\ \vdots & \ddots & \vdots & \vdots \\ 0 & \cdots & I_n & 0 \end{pmatrix} x_{t-1}^p + \begin{pmatrix} I_n \\ 0 \\ \vdots \\ 0 \end{pmatrix} \epsilon_t$$

$$x_t = \begin{pmatrix} a_1 & a_2 & \cdots & a_p \end{pmatrix} x_{t-1}^p + \epsilon_t.$$

Exercise 7.19 Prove that the following statements are equivalent for a regular process (x_t):

(1) (x_t) is an AR(p) process.
(2) $\mathbb{H}_t^+(x) \subset \mathrm{sp}\{x_{t-1}, \ldots, x_{t-p}\}$.
(3) $(A - BC)^p = 0$ holds for a minimal state-space representation in innovations form.
(4) The transfer function (corresponding to the Wold representation) is rational and has *no* zeros.

Note: Lemma 7.9 can be applied for the equivalence of (3) and (4).

Exercise 7.20 Let $(x_t = \epsilon_t + b_1\epsilon_{t-1} + \cdots + b_q\epsilon_{t-q})$ be an MA(q) process. Verify the following state-space representation for the process (x_t):

$$\epsilon_t^q = \begin{pmatrix} 0 & \cdots & 0 & 0 \\ I_n & \cdots & 0 & 0 \\ \vdots & \ddots & \vdots & \vdots \\ 0 & \cdots & I_n & 0 \end{pmatrix} \epsilon_{t-1}^q + \begin{pmatrix} I_n \\ 0 \\ \vdots \\ 0 \end{pmatrix} \epsilon_t$$

$$x_t = \begin{pmatrix} b_1 & b_2 & \cdots & b_q \end{pmatrix} \epsilon_{t-1}^q + \epsilon_t.$$

Exercise 7.21 Prove that the following statements are equivalent for a regular process (x_t):

(1) (x_t) is an MA(q) process.
(2) $\mathbb{H}_t^+(x) \subset \mathrm{sp}\{\epsilon_{t-1}, \ldots, \epsilon_{t-q}\}$, where the ϵ_t's are the innovations of (x_t).
(3) $A^q = 0$ holds for a minimal state-space representation in innovations form.
(4) The transfer function (corresponding to the Wold representation) is rational and has no poles.

Note: Lemma 7.6 can be applied for the equivalence of (3) and (4).

7.4   The Kalman Filter

The *Kalman filter* is a recursive method to estimate the unobserved state of a (linear, dynamic) state-space system from observations of the output of the system. The filter is named after Rudolf E. Kálmán, who has made important contributions to systems theory and the development of this filter. The Kalman filter is widely applied in technical fields such as navigation, control, monitoring and signal processing. In time series analysis and econometrics, the Kalman filter is mainly used for prediction and for maximum likelihood estimation of state-space systems. It is also suitable for real-time applications due to its recursive structure.

The underlying model is a state-space system of the form

$$s_{t+1} = As_t + Ev_t + \xi_t \tag{7.27}$$
$$x_t = Cs_t + \eta_t \tag{7.28}$$

which is started at time $t = 1$ with the initial value s_1. We assume that the input process (v_t) is non-stochastic and known for all $t \in \mathbb{N}$ and that the output process (x_t) is observed. The state process (s_t) and the disturbances (ξ_t), (η_t) are latent processes, i.e. they are unobserved. The stacked error process $(\xi_t', \eta_t')'$ is white noise uncorrelated to the initial state s_1. That is, we require for $t, r \in \mathbb{N}$

$$
\begin{array}{lll}
\mathbf{E}\xi_t = 0 & \mathbf{E}\eta_t = 0 & \mathbf{E}s_1 = s_{1|0} \\
\mathbf{E}\xi_r\xi_t' = \delta_{rt}Q & \mathbf{E}\xi_r\eta_t' = \delta_{rt}S & \mathbf{E}\eta_r\eta_t' = \delta_{rt}R \\
\mathbf{E}s_1\xi_t' = 0 & \mathbf{E}s_1\eta_t' = 0 & \mathbf{Var}(s_1) = \Pi_{1|0}
\end{array}
$$

where δ_{rt} stands for the Kronecker delta. The system matrices (A, E, C), the covariance matrices $(Q, R, S, \Pi_{1|0})$ and the expectation $s_{1|0}$ are assumed to be known.

This model is more general than the state-space model discussed previously in Sects. (7.1), (7.2). In particular, there are two noise sources, (ξ_t) in the state equation and (η_t) in the observation equation and one non-stochastic, observed input (v_t). Apart from the structure of the model, we do not impose further conditions, such as stability or minimality of the system.

The Kalman filter is a recursive (in t) method, that is used to compute the optimal (affine) least squares estimates for future states s_{t+h} and outputs x_{t+h} ($h \geq 0$) from the observations $x_1, \ldots, x_t$. According to the projection theorem, these least squares estimators are obtained by projection onto the Hilbert space $\mathbb{H}_{1:t}(x) := \mathrm{sp}\{1, x_1, \ldots, x_t\}$. We denote the corresponding projection operator by $P_t = P_{\mathbb{H}_{1:t}(x)}$ and for the projections of the states and outputs onto $\mathbb{H}_{1:t}(x)$, we introduce the following notation:

$$
\begin{array}{llll}
P_t s_r = s_{r|t} & & \mathbf{Var}(s_r - s_{r|t}) = \Pi_{r|t} \\
P_t x_r = x_{r|t} & u_{r|t} = x_r - x_{r|t} & \mathbf{Var}(u_{r|t}) = \Sigma_{r|t}.
\end{array}
$$

By writing $P_0 = P_{\mathrm{sp}\{1\}}$, we have $s_{1|0} = \mathbf{E}s_1 = P_0 s_1$ and $\Pi_{1|0} = \mathbf{Var}(s_1) = \mathbf{Var}(s_1 - s_{1|0})$. The notations $s_{1|0}$ and $\Pi_{1|0}$ are thus consistent with the above conventions.

Proposition 7.22 (Kalman filter) *Under the assumptions given above, the one-step ahead predictions are calculated by the following recursive system (for $t > 0$):*

$$u_{t|t-1} = x_t - x_{t|t-1}$$
$$K_t = (A\Pi_{t|t-1}C' + S)\Sigma_{t|t-1}^{-1}$$
$$s_{t+1|t} = As_{t|t-1} + Ev_t + K_t u_{t|t-1}$$
$$\Pi_{t+1|t} = \mathbf{Var}(s_{t+1} - s_{t+1|t}) = A\Pi_{t|t-1}A' + Q - K_t\Sigma_{t|t-1}K_t'$$
$$x_{t+1|t} = Cs_{t+1|t}$$
$$\Sigma_{t+1|t} = \mathbf{Var}(u_{t+1|t}) = C\Pi_{t+1|t}C' + R.$$

This system is initialized with $s_{1|0}$, $\Pi_{1|0}$, $x_{1|0} = Cs_{1|0}$ and $\Sigma_{1|0} = C\Pi_{1|0}C' + R$. For the h-step ahead prediction ($h > 1$), we have

$$s_{t+h|t} = As_{t+h-1|t} + Ev_{t+h-1}$$
$$\Pi_{t+h|t} = \mathbf{Var}(s_{t+h} - s_{t+h|t}) = A\Pi_{t+h-1|t}A' + Q$$
$$x_{t+h|t} = Cs_{t+h|t}$$
$$\Sigma_{t+\bar{h}|t} = \mathbf{Var}(x_{t+h} - x_{t+h|t}) = C\Pi_{t+h|t}C' + R$$

and for $h = 0$:

$$s_{t|t} = s_{t|t-1} + \Pi_{t|t-1}C'\Sigma_{t|t-1}^{-1}u_{t|t-1}$$
$$\Pi_{t|t} = \mathbf{Var}(s_t - s_{t|t}) = \Pi_{t|t-1} - \Pi_{t|t-1}C'\Sigma_{t|t-1}^{-1}C\Pi_{t|t-1}.$$

Proof The whole proof is based on using certain orthogonality relations in a clever way to determine the projections. In particular:

$$(\xi'_{t+h}, \eta'_{t+h})' \perp \text{sp}\{1, s_1, \ldots, s_{t+1}, x_1, \ldots, x_t\} \; \forall h > 0 \qquad (7.29)$$

since

$$\mathbb{H}_{1:t}(x) \subset \text{sp}\{1, s_1, \ldots, s_t, s_{t+1}, x_1, \ldots, x_t\} \subset \text{sp}\{1, s_1, \xi_1, \ldots, \xi_t, \eta_1, \ldots, \eta_t\}.$$

We will also repeatedly use the following elementary properties of the projection: Let u, v be two random vectors and $P = P_{\mathbb{H}}$ be the projection onto a Hilbert space $\mathbb{H}$; then it is easy to show that

$$E(u - Pu)(u - Pu)' = Euu' - E(Pu)(Pu)' \qquad (7.30)$$
$$Eu(v - Pv)' = E(u - Pu)(v - Pv)'. \qquad (7.31)$$

$\square$

If $\mathbb{H} = \mathbb{H}_1 \oplus \mathbb{H}_2$ is the sum of two orthogonal spaces and hence $P_\mathbb{H} = P_{\mathbb{H}_1} + P_{\mathbb{H}_2}$, then it also follows that

$$\mathbf{E}(u - P_\mathbb{H}\,u)(u - P_\mathbb{H}\,u)' = \mathbf{E}(u - P_{\mathbb{H}_1}\,u)(u - P_{\mathbb{H}_1}\,u)' - \mathbf{E}(P_{\mathbb{H}_2}\,u)(P_{\mathbb{H}_2}\,u)'. \tag{7.32}$$

If the projection $s_{t+h|t}$, $h > 0$ is known, then it is very simple to compute the prediction $x_{t+h|t}$:

$$
\begin{aligned}
x_{t+h|t} &= P_t(Cs_{t+h} + \eta_{t+h}) = Cs_{t+h|t} \\
u_{t+h|t} &= x_{t+h} - x_{t+h|t} = C(s_{t+h} - s_{t+h|t}) + \eta_{t+h} \\
\Sigma_{t+h|t} &= \mathbf{E}u_{t+h|t}u'_{t+h|t} = \mathbf{E}(C(s_{t+h} - s_{t+h|t}) + \eta_{t+h})(C(s_{t+h} - s_{t+h|t}) + \eta_{t+h})' \\
&= C\Pi_{t+h|t}C' + R.
\end{aligned}
$$

Here we have used $\eta_{t+h} \perp \mathbb{H}_{1:t}(x)$ (and thus $P_t\,\eta_{t+h} = 0$ and $\eta_{t+h} \perp s_{t+h|t}$) and $\eta_{t+h} \perp s_{t+h}$.

The one-step ahead prediction errors $u_{r|r-1}, r = 1, \ldots, t$, together with the constant 1, span the Hilbert space $\mathbb{H}_{1:t}(x)$ and are mutually orthogonal. Hence, it follows directly that

$$
\begin{aligned}
\mathbb{H}_{1:t}(x) &= \mathbb{H}_{1:t-1}(x) \oplus \mathrm{sp}\{u_{t|t-1}\} = \mathrm{sp}\{1\} \oplus \mathrm{sp}\{u_{1|0}\} \oplus \cdots \oplus \mathrm{sp}\{u_{t|t-1}\} \\
P_t &= P_{t-1} + P_t^u = P_0 + P_1^u + \cdots + P_t^u
\end{aligned}
$$

where $P_r^u := P_{\mathrm{sp}\{u_{r|r-1}\}}$. This allows us to obtain the following recursive equations for the estimation of the states:

$$
\begin{aligned}
s_{t+1|t} = P_t\,s_{t+1} &= P_{t-1}(As_t + Ev_t + \xi_t) + P_t^u\,s_{t+1} \\
&= As_{t|t-1} + Ev_t + K_t u_{t|t-1}
\end{aligned}
$$

where $K_t = \mathbf{E}(s_{t+1}u'_{t|t-1})\mathbf{E}(u_{t|t-1}u'_{t|t-1})^{-1}$ denotes the so-called *Kalman gain*. The projection $P_{t-1}\,\xi_t$ is equal zero because $\xi_t \perp \mathbb{H}_{1:t-1}(x)$. We have

$$
\begin{aligned}
\mathbf{E}(s_{t+1}u'_{t|t-1}) &= \mathbf{E}(As_t + Ev_t + \xi_t)(C(s_t - s_{t|t-1}) + \eta_t)' = \\
&= A\mathbf{E}\left[(s_t - s_{t|t-1})(s_t - s_{t|t-1})'\right]C' + \mathbf{E}\xi_t\eta_t' + \\
&\quad + \mathbf{E}\left[Ev_t((s_t - s_{t|t-1})'C' + \eta_t')\right] + \left[\mathbf{E}\xi_t(s_t - s_{t|t-1})'\right]C' + AEs_t\eta_t' \\
&= A\Pi_{t|t-1}C' + S
\end{aligned}
$$

and hence

$$K_t = (A\Pi_{t|t-1}C' + S)\Sigma_{t|t-1}^{-1}.$$

The terms in the third row above are zero since v_t is not stochastic, $(s_t - s_{t|t-1})$ and η_t have expectation zero and because $(\xi_t', \eta_t')' \perp s_t$ and $\xi_t \perp \mathbb{H}_{1:t-1}(x)$. For the variance of the approximation error $(s_{t+1} - s_{t+1|t})$, we obtain with the help of (7.32)

$$\Pi_{t+1|t} = \mathbf{E}(s_{t+1} - P_{t-1}\, s_{t+1})(s_{t+1} - P_{t-1}\, s_{t+1})' - \mathbf{E}(P_t^u\, s_{t+1})(P_t^u\, s_{t+1})'$$
$$= \mathbf{E}(A(s_t - s_{t|t-1}) + \xi_t)(A(s_t - s_{t|t-1}) + \xi_t)' - K_t \Sigma_{t|t-1} K_t'$$
$$= A\Pi_{t|t-1} A' + Q - K_t \Sigma_{t|t-1} K_t'.$$

For $h > 1$,

$$s_{t+h|t} = P_t\, s_{t+h} = P_t(As_{t+h-1} + Ev_{t+h-1} + \xi_{t+h-1})$$
$$= A\, P_t\, s_{t+h-1} + Ev_{t+h-1} = As_{t+h-1|t} + Ev_{t+h-1}$$

$$\Pi_{t+h|t} = \mathbf{E}(s_{t+h} - s_{t+h|t})(s_{t+h} - s_{t+h|t})$$
$$= \mathbf{E}(A(s_{t+h-1} - s_{t+h-1|t}) + \xi_{t+h-1})(A(s_{t+h-1} - s_{t+h-1|t}) + \xi_{t+h-1})'$$
$$= A\Pi_{t+h-1|t} A' + Q.$$

Now we consider the case $h = 0$, i.e. we want to derive the formulas for $s_{t|t}$ and $\Pi_{t|t}$:

$$s_{t|t} = P_{t-1}\, s_t + P_t^u\, s_t = s_{t|t-1} + M_t u_{t|t-1}$$

$$\Pi_{t|t} = \mathbf{E}(s_t - s_{t|t-1})(s_t - s_{t|t-1})' - \mathbf{E}(P_t^u\, s_t)(P_t^u\, s_t)' = \Pi_{t|t-1} - M_t \Sigma_{t|t-1} M_t$$

where

$$M_t \Sigma_{t|t-1} = \mathbf{E}(s_t u_{t|t-1}') = \mathbf{E}\left[s_t((s_t - s_{t|t-1})'C' + \eta_t')\right]$$
$$= \mathbf{E}\left[(s_t - s_{t|t-1})'(s_t - s_{t|t-1})'\right] C' = \Pi_{t|t-1} C'.$$

The Kalman filter gives the (affine) least squares approximation of the state s_t from past observations $x_1, \ldots, x_r, r \leq t$. The next step is now to estimate the state s_t from past *and* future observations $x_1, \ldots, x_r, r \geq t$. This is called *smoothing*. This problem can also be solved elegantly by a recursive procedure.

Proposition 7.23 (Kalman smoothing) *For* $1 < t \leq r$:

$$s_{t-1|r} = s_{t-1|t-1} + J_{t-1}(s_{t|r} - s_{t|t-1})$$
$$\Pi_{t-1|r} = \Pi_{t-1|t-1} + J_{t-1}(\Pi_{t|r} - \Pi_{t|t-1})J_{t-1}'$$

where

$$J_{t-1} = \Pi_{t-1|t-2}(A' - C'K_{t-1}')\Pi_{t|t-1}^{-1}.$$

To compute these estimators $s_{t|r}$, one first determines in a forward recursion for $t = 1, 2, \ldots, r$ using the Kalman filter $s_{t|t}$, $\Pi_{t|t}$, $s_{t|t-1}$, $\Pi_{t|t-1}$ and K_t. Then one uses the above equations in a backward recursion for $t = r - 1, r - 2, \ldots, 1$ to compute $s_{t|r}$ and $\Pi_{t|r}$.

Proof The proof given here goes back to Ansley and Kohn (1982). The Hilbert space $\mathbb{H}_{1:r}(x)$ is a subspace of

$$\mathbb{H}(r) := \mathbb{H}_{1:t-1}(x) \oplus \mathrm{sp}\{s_t - s_{t|t-1}\} \oplus \mathrm{sp}\{\eta_t, \ldots, \eta_r, \xi_t, \ldots, \xi_{r-1}\}.$$

Considering that the space $\mathbb{H}(r)$ is a direct sum of three mutually orthogonal spaces, the projection of s_{t-1} onto $\mathbb{H}(r)$ is given by

$$P_{\mathbb{H}(r)} s_{t-1} = P_{t-1} s_{t-1} + J_{t-1}(s_t - s_{t|t-1}) + 0,$$

where $J_{t-1}(s_t - s_{t|t-1})$ is the projection of s_{t-1} onto the space $\mathrm{sp}\{s_t - s_{t|t-1}\}$. Here we have also used $(\xi'_{t+h}, \eta'_{t+h}) \perp s_{t-1}$ for $h \geq 0$. The matrix J_{t-1} is computed from

$$J_{t-1} = \mathbf{E}\left[s_{t-1}(s_t - s_{t|t-1})'\right]\left(\mathbf{E}\left[(s_t - s_{t|t-1})(s_t - s_{t|t-1})'\right]\right)^{-1} =$$

$$\mathbf{E}\left[s_{t-1}(A(s_{t-1} - s_{t-1|t-2}) + \xi_{t-1} - K_{t-1}(C(s_{t-1} - s_{t-1|t-2}) + \eta_{t-1}))'\right]\Pi_{t|t-1}^{-1} =$$

$$\Pi_{t-1|t-2}(A' - C'K'_{t-1})\Pi_{t|t-1}^{-1}.$$

Since $\mathbb{H}_{1:r}(x) \subset \mathbb{H}(r)$, it now follows that

$$s_{t-1|r} = P_r\, s_{t-1} = P_r\, P_{\mathbb{H}(r)}\, s_{t-1} = P_r\, P_{t-1}\, s_{t-1} + J_{t-1}(P_r\, s_t - P_r\, s_{t|t-1})$$
$$= s_{t-1|t-1} + J_{t-1}(s_{t|r} - s_{t|t-1}).$$

The formula for the variances of the error of estimation is obtained with the help of (7.32) and the relation $(s_{t-1|r} - s_{t-1|t-1}) = J_{t-1}(s_{t|r} - s_{t|t-1})$:

$$\Pi_{t-1|t-1} - \Pi_{t-1|r} = \mathbf{Var}(s_{t-1|r} - s_{t-1|t-1}) = J_{t-1}\mathbf{Var}(s_{t|r} - s_{t|t-1})J'_{t-1}$$
$$= J_{t-1}(\Pi_{t|t-1} - \Pi_{t|r})J'_{t-1}.$$

$\square$

In the recursive equations for the Kalman filter and Kalman smoother, the inverses of the covariance matrices $\Sigma_{t|t-1}$ and $\Pi_{t|t-1}$ appear. If these matrices are singular, then one may use the Moore–Penrose inverse instead. (The Moore–Penrose inverse of a matrix is obtained from the singular value decomposition of the matrix by replacing all nonzero singular values with their reciprocals.) The Kalman gain K_t, for example, is defined by the projection of s_{t+1} onto the space $\text{sp}\{u_{t|t-1}\}$. That is, K_t is a solution of the equation

$$\underbrace{\mathbf{E}(s_{t+1}u'_{t|t-1})}_{=(A\Pi_{t|t-1}C'+S)} = K_t \underbrace{\mathbf{E}(u_{t|t-1}u'_{t|t-1})}_{=\Sigma_{t|t-1}}.$$

This equation can always be solved (due to the projection theorem) and every solution K_t yields the same projection $\mathbf{P}_t^u s_{t+1} = K_t u_{t|t-1}$. In particular, $K_t = (A\Pi_{t|t-1}C' + S)\Sigma_{t|t-1}^{\dagger}$, where $\Sigma_{t|t-1}^{\dagger}$ denotes the Moore–Penrose inverse, is a solution, because the column space of $\mathbf{E}(s_{t+1}u'_{t|t-1})$ is a subspace of the column space of $\mathbf{E}(u_{t|t-1}u'_{t|t-1})$. Analogous considerations apply to the computation of $s_{t|t}$ and the computation of the matrix J_{t-1}. See also the Exercise 1.13 on Sect. 1.3.

The Kalman filter and Kalman smoother have been formulated here for the case of *constant* parameters (A, E, C, Q, R, S). However, the results can be generalized quite easily to a state-space model

$$\begin{aligned} s_{t+1} &= A_t s_t + E_t v_t + \xi_t \\ x_t &= C_t s_t + \eta_t \end{aligned} \quad \text{and } \mathbf{E}\begin{pmatrix}\xi_r \\ \eta_r\end{pmatrix}\begin{pmatrix}\xi_t \\ \eta_t\end{pmatrix}' = \delta_{rt}\begin{pmatrix}Q_t & S_t \\ S'_t & R_t\end{pmatrix}$$

with time-dependent parameters. The essential recursive equations for the Kalman filter then are, e.g.,

$$K_t = (A_t\Pi_{t|t-1}C'_t + S_t)\Sigma_{t|t-1}^{-1}$$
$$s_{t+1|t} = A_t s_{t|t-1} + E_t v_t + K_t u_{t|t-1}$$
$$\Pi_{t+1|t} = A_t\Pi_{t|t-1}A'_t + Q_t - K_t\Sigma_{t|t-1}K'_t$$
$$x_{t+1|t} = C_{t+1}s_{t+1|t}$$
$$\Sigma_{t+1|t} = C_{t+1}\Pi_{t+1|t}C'_{t+1} + R_{t+1}.$$

Missing observations can be considered without major problems. As a simple example, we consider the case of a model with constant parameters and assume that x_{t_0} was not observed. The missing observation x_{t_0} is now modeled by making the corresponding covariance matrix of the perturbations η_{t_0} very large, that is, we write $R_{t_0} = \mathbf{E}\eta_{t_0}\eta'_{t_0} = R + cI_n$ and consider the Kalman filter for the limit for $c \to \infty$. If the covariance matrix R_{t_0} is very large, then x_{t_0} contains little information about the underlying state s_{t_0} and in the limiting case $R_{t_0} \to \infty I$, x_{t_0} contains no information and it does not matter whether x_{t_0} is observed or not. (Alternatively, we could also consider the case $C_{t_0} = 0$ and $S_{t_0} = 0$.)

$$\Sigma_{t_0|t_0-1} = C\Pi_{t_0|t_0-1}C' + R_{t_0} \longrightarrow \infty I_n$$
$$K_{t_0} = (A\Pi_{t_0|t_0-1}C' + S)\Sigma_{t_0|t_0-1}^{-1} \longrightarrow 0$$
$$s_{t_0+1|t_0} = As_{t_0|t_0-1} + Ev_{t_0} + K_{t_0}u_{t_0|t_0-1} \longrightarrow As_{t_0|t_0-1} + Ev_{t_0} + 0 = s_{t_0+1|t_0-1}$$
$$\Pi_{t_0+1|t_0} = A\Pi_{t_0|t_0-1}A' + Q - K_{t_0}\Sigma_{t_0|t_0-1}K'_{t_0}$$

$$\longrightarrow A\Pi_{t_0|t_0-1}A' + Q - 0 = \Pi_{t_0+1|t_0-1}$$

$$x_{t_0+1|t_0} = Cs_{t_0+1|t_0} \longrightarrow x_{t_0+1|t_0-1}$$

$$\Sigma_{t_0+1|t_0} = C\Pi_{t_0+1|t_0}C' + R \longrightarrow \Sigma_{t_0+1|t_0-1}.$$

The Kalman filter here simply "skips" the time t_0, i.e. one-step ahead predictions $x_{t_0+1|t_0}$ and $s_{t_0+1|t_0}$ are replaced by the two-step ahead predictions $x_{t_0+1|t_0-1}$ and $s_{t_0+1|t_0-1}$. The appropriately adapted Kalman smoother can be used to compute an estimate for the unobserved value x_{t_0} from the available observations $\{x_t, 1 \le t \le r, t \ne t_0\}$. According to this scheme, one can also handle more complicated scenarios for missing observations, e.g. the case where only some components of x_{t_0} are missing.

For the Kalman filter and Kalman smoother, it is not assumed that the state process (s_t) and the output process (x_t) are stationary. (However, since we are computing affine least squares approximations, s_t and x_t must be, of course, square integrable.) Thus, the transition matrix A needs not to be stable; it may have eigenvalues with magnitude one or greater than one. Of course, non-stationary processes will also occur in the case of time-dependent parameters.

Exercise 7.24 (*simple exponential smoothing*) We consider the state-space model

$$\begin{aligned} s_{t+1} &= s_t + \xi_t \\ x_t &= s_t + \eta_t \end{aligned}, \quad \mathbf{E}\begin{pmatrix} \xi_r \\ \eta_r \end{pmatrix}\begin{pmatrix} \xi_t \\ \eta_t \end{pmatrix}' = \delta_{rt}\begin{pmatrix} Q & 0 \\ 0 & R \end{pmatrix}$$

with scalar states s_t and outputs x_t. The state process (s_t) is a random walk process and the output (x_t) is therefore a random walk process superposed by white noise. Show that the Kalman filter is of the form

$$s_{t+1|t} = s_{t|t-1} + Ku_{t|t-1} = (1 - K)s_{t|t-1} + Kx_t$$

$$x_{t+h|t} = s_{t+1|t}$$

if the filter is initialized with

$$\Pi_{1|0} = \frac{Q + \sqrt{Q^2 - 4RQ}}{2}.$$

The above recursive equations correspond to the so-called *simple exponential smoothing* with smoothing factor $K = \frac{\Pi_{1|0}}{\Pi_{1|0}+R}$. Exponential smoothing is a simple heuristic prediction method.

The initial value $s_{1|0}$ is a conjecture about the unknown initial state s_1 and the covariance matrix $\Pi_{1|0}$ reflects the confidence in this conjecture. If a realistic conjecture about s_1 is not possible, then one often sets $s_{1|0} = 0$ and chooses $\Pi_{1|0} = cI_m$ with a very large c. This heuristic procedure can be formalized with the help of the so-called *diffuse Kalman filter*. Here we write $\Pi_{1|0} = \Pi_{1|0}^0 + c\Pi_{1|0}^\infty$ and then analyze the Kalman filter recursions for the limit $c \longrightarrow \infty$.

Exercise 7.25 (*Recursive Least Squares (RLS) Estimator*) We consider a classical regression model $x_t = v_t\beta + u_t$, with deterministic regressors $v_t \in \mathbb{R}^{1 \times m}$ and homoscedastic and uncorrelated errors (i.e. $\mathbf{E}u_t = 0$, $\mathbf{E}u_r u_t = \delta_{rt}\sigma^2$). The ordinary least squares (OLS) estimator for β from observations $(x_1, \ldots, x_t)$ is

$$\hat{\beta}_t = \left(\sum_{i=1}^t v_i' v_i\right)^{-1} \left(\sum_{i=1}^t v_i' x_i\right) \text{ and } \mathbf{Var}(\hat{\beta}_t) = \sigma^2 \left(\sum_{i=1}^t v_i' v_i\right)^{-1}.$$

The assumption here is, of course, that the matrix $\left(\sum_{i=1}^t v_i' v_i\right)$ is non-singular. You are now supposed to show that the Kalman filter is a possibility to recursively (in t) determine the estimator. For this purpose, we first write the regression model as a state-space model

$$\begin{matrix} s_{t+1} = s_t + \xi_t \\ x_t = v_t s_t + u_t \end{matrix} \; ; \; \mathbf{E}\begin{pmatrix} \xi_r \\ u_r \end{pmatrix}\begin{pmatrix} \xi_t \\ u_t \end{pmatrix}' = \delta_{rt}\begin{pmatrix} 0 & 0 \\ 0 & \sigma^2 \end{pmatrix}$$

where the state $s_t = \beta$ is the constant coefficient vector β, to be estimated. Since β does not depend on t, we set $\xi_t = 0$.

Let $t_0 \geq m$ be the smallest time index for which $\left(\sum_{i=1}^t v_i' v_i\right)$ is non-singular. Show that the Kalman filter yields the desired recursive estimator, i.e., show

$$s_{t+1|t} = \hat{\beta}_t$$

$$\Pi_{t+1|t} = \frac{1}{\sigma^2}\mathbf{Var}(\hat{\beta}_t)$$

for $t > t_0$, if one initializes the filter at time t_0 with $s_{t_0+1|t_0} = \hat{\beta}_{t_0}$ and $\Pi_{t_0+1|t_0} = \frac{1}{\sigma^2}\mathbf{Var}(\hat{\beta}_{t_0}) = \left(\sum_{i=1}^{t_0} v_i' v_i\right)^{-1}$. In this example, it is therefore "clear" how the initial values of the filter must be chosen.

If an adaptive estimation of time-dependent coefficients β_t is desired, then we simply set, e.g., $Q = \mathbf{E}\xi_t \xi_t' = \epsilon I_m$. The constant $\epsilon > 0$ then controls the "adaptivity" or the response speed of the estimator, respectively: For small ϵ, the estimator $s_{t+1|t}$ responds in a relatively slow manner to changes in the coefficient vector and therefore a relatively "smooth" response is obtained.

The Kalman filter is a very popular algorithm, used in numerous technical and scientific fields, such as control, signal processing, forecasting, etc. Hence, a whole range of alternative implementations and extensions exist, see e.g. Anderson and Moore (2005), of which only a few shall be mentioned here. The diffuse Kalman filter, which handles the case of (partly) unknown initial values $s_{1|0}$ and $\Pi_{1|0}$, has already been briefly mentioned above. In the information filter, one uses recursive equations for the inverse covariance matrix $I_{t|t-1} = \Pi_{t|t-1}^{-1}$. So-called "square root" filters use recursions for the square roots $\Pi_{t|t-1}^{1/2}$ of the covariance matrices. These filters have numerical advantages.

We will now discuss the stationary case in more detail. For the sake of simplicity, we omit the input process (v_t) and write $E = 0$. Moreover, we always assume the stability condition $\varrho(A) < 1$ and consider the (unique) stationary solution

$$x_t = \sum_{j \geq 0} CA^j \xi_{t-1-j} + \eta_t \tag{7.33}$$

of the state-space system (7.27) and (7.28).

Proposition 7.26 *The process (7.33) has a state-space representation of the form*

$$\tilde{s}_{t+1} = A\tilde{s}_t + B\epsilon_t \tag{7.34}$$
$$x_t = C\tilde{s}_t + \epsilon_t, \quad t \in \mathbb{Z} \tag{7.35}$$

where $C\tilde{s}_t$ is the one-step ahead predictor of x_t from the infinite past ($\mathbb{H}_{t-1}(x)$) and the ϵ_t's are the innovations of (x_t).

Proof We project (7.27) onto the Hilbert space $\mathbb{H}_t(x)$ generated by $\{x_r \mid r \leq t\}$. Then we obtain (in evident notation):

$$\hat{s}_{t+1,t} = A\hat{s}_{t,t} + \hat{\xi}_{t,t} = A\hat{s}_{t,t-1} + \left(A(\hat{s}_{t,t} - \hat{s}_{t,t-1}) + \hat{\xi}_{t,t} \right)$$

and analogously for (7.28)

$$x_t = \hat{x}_{t,t} = C\hat{s}_{t,t-1} + \left(C(\hat{s}_{t,t} - \hat{s}_{t,t-1}) + \hat{\eta}_{t,t} \right).$$

Now we define $\epsilon_t = \left(C(\hat{s}_{t,t} - \hat{s}_{t,t-1}) + \hat{\eta}_{t,t} \right)$. The random vector ϵ_t is orthogonal to $\mathbb{H}_{t-1}(x)$, since for $r < t$

$$\mathbf{E}\epsilon_t x_r' = C(\mathbf{E}\hat{s}_{t,t} x_r' - \mathbf{E}\hat{s}_{t,t-1} x_r') + \mathbf{E}\hat{\eta}_{t,t} x_r' = C(\mathbf{E}s_t x_r' - \mathbf{E}s_t x_r') + \mathbf{E}\eta_t x_r' = 0.$$

Thus, $C\hat{s}_{t,t-1} \in (\mathbb{H}_{t-1}(x))^n$ is the one-step ahead predictor and ϵ_t is the associated prediction error. In particular, ϵ_t therefore spans the space $\mathbb{H}_t(x) \ominus \mathbb{H}_{t-1}(x) = \overline{sp}\{u \mid u \in \mathbb{H}_t(x) \text{ and } u \perp \mathbb{H}_{t-1}(x)\}$, i.e. the orthogonal complement of $\mathbb{H}_{t-1}(x)$ in $\mathbb{H}_t(x)$. Since $\left(A(\hat{s}_{t,t} - \hat{s}_{t,t-1}) + \hat{\xi}_{t,t} \right)$ lies in this space, there exists a matrix B such that $\left(A(\hat{s}_{t,t} - \hat{s}_{t,t-1}) + \hat{\xi}_{t,t} \right) = B\epsilon_t$. We now write $\tilde{s}_t = \hat{s}_{t,t-1}$ and hence obtain (7.34) and (7.35).   □

If the system (7.34), (7.35) is minimal, then it is in innovations form, i.e. the stability condition $\varrho(A) < 1$ and the minimum-phase condition $\varrho(A - BC) < 1$ hold. Of course, conversely, one can also write the model (7.34), (7.35) in the form (7.27), (7.28) by $\xi_t = B\epsilon_t$, $\eta_t = \epsilon_t$ and hence $Q = B\Sigma B'$, $S = B\Sigma$ and $R = \Sigma$.

In this case, the correct initialization of the filter is

$$s_{1|0} = \mathbf{E}s_1 = 0 \ \text{ and } \ P := \Pi_{1|0} = \mathbf{E}s_1 s_1' = \sum_{j\geq 0} A^j Q (A^j)'.$$

The variance P of the state s_t can also be determined by solving the Lyapunov equation

$$P = APA' + Q.$$

The Kalman filter computes the projection of state s_{t+1} and of the future output x_{t+1} onto the space $\text{sp}\{x_1, \ldots, x_t\}$. (Since $\mathbf{E}s_t = 0$ and $\mathbf{E}x_t = 0$, it suffices to consider linear approximations here; we can therefore omit the constant "1".) Hence it follows immediately that

$$\text{l.i.m}_{t\to\infty}(s_{t+1|t} - \tilde{s}_{t+1}) = 0 \ \text{ and } \ \text{l.i.m}_{t\to\infty}(u_{t+1|t} - \epsilon_{t+1}) = 0.$$

Exercise 7.27 We commence from a system of the form (7.27) and (7.28) without an external input v_t and a stable state transition matrix A. The corresponding system in innovations form is (7.34) and (7.35). We define $\Sigma = \mathbf{E}\epsilon_t \epsilon_t'$ and $\tilde{P} = \mathbf{E}\tilde{s}_t \tilde{s}_t' = A\tilde{P}A' + B\Sigma B'$. Apply the Kalman filter to the system (7.27) and (7.28). Convince yourself that

$$\Sigma_{t|t-1} \longrightarrow \Sigma, \ \Pi_{t|t-1} \longrightarrow P - \tilde{P} \ \text{ and } \ K_t \Sigma_{t|t-1} = (A\Pi_{t|t-1}C' + S) \longrightarrow B\Sigma.$$

Thus, the Kalman filter recursions for the one-step ahead errors converge (under the condition $\Sigma > 0$) for $t \to \infty$ to the recursion equations of the system inverse of (7.34), (7.35):

$$\begin{array}{ll}
s_{t+1|t} = (A - K_t C)s_{t|t-1} + K_t u_{t|t-1} & \xrightarrow{t\to\infty} \quad \tilde{s}_{t+1} = (A - BC)\tilde{s}_t + B\epsilon_t \\
u_{t|t-1} = -Cs_{t|t-1} + x_t & \qquad\qquad \epsilon_t = -C\tilde{s}_t + x_t.
\end{array}$$

The following two exercises show that the autocovariance function of the stationary output (x_t) can be computed in a fairly simple way:

Exercise 7.28 Consider a state-space system (7.27), (7.28) with white noise $((\xi_t', \eta_t')')$ and no exogenous inputs (v_t), that satisfies the stability condition $\varrho(A) < 1$. Show that the covariance function γ of the stationary solution (x_t), see (7.33), is given by

$$\gamma(0) = CPC' + R$$
$$\gamma(k) = CA^{k-1}M, \ \text{ for } k > 0$$

where

$$P = \mathbf{E}s_t s_t' = APA' + Q$$
$$M = \mathbf{E}s_{t+1}x_t = APC' + S.$$

Exercise 7.29 (*extension*) With the help of the equivalent state-space representation (7.34), (7.35), derive the following alternative representation of the covariance function γ:

$$\gamma(0) = C\tilde{P}C' + \Sigma$$
$$\gamma(k) = CA^{k-1}\tilde{M}, \quad \text{for } k > 0$$

where

$$\tilde{P} = \mathbf{E}\tilde{s}_t\tilde{s}_t' = A\tilde{P}A' + B\Sigma B'$$
$$\tilde{M} = \mathbf{E}\tilde{s}_{t+1}x_t = A\tilde{P}C' + B\Sigma.$$

References

H. Akaike, A new look at the statistical model identification. IEEE Trans. Autom. Control **19**(6), 716–723 (1974). ISSN 0018-9286. 10.1109/TAC.1974.1100705

B.D.O. Anderson, J.B. Moore, *Optimal Filtering* (Dover Publications Inc., London, 2005). (Originally published: Englewood Cliffs, Prentice-Hall 1979)

C.F. Ansley, R. Kohn, A geometrical derivation of the fixed interval smoothing algorithm. Biometrika **69**(2), 486–487 (1982). ISSN 00063444. http://www.jstor.org/stable/2335428

V. Gómez, *Multivariate Time Series with Linear State Space Structure* (Springer, 2016). ISBN 978-3-319-28598-6; 3-319-28598-X

E.J. Hannan, M. Deistler, *The Statistical Theory of Linear Systems*. Classics in Applied Mathematics (SIAM, Philadelphia, 2012). (Originally published: Wiley, New York, 1988)

T. Kailath, *Linear Systems* (Prentice Hall, Englewood Cliffs, New Jersey, 1980)

R.E. Kalman, A new approach to linear filtering and prediction problems. Trans. ASME. J. Basic Eng. **82**, 35–45 (1960)

R.E. Kalman, Mathematical description of linear dynamical systems. J. Soc. Ind. Appl. Math. Ser. Control **1**(2), 152–192 (1963). https://doi.org/10.1137/0301010

R.E. Kalman, Irreducible realizations and the degree of a rational matrix. J. Soc. Ind. Appl. Math. **13**(2), 520–544 (1965). https://doi.org/10.1137/0113034

R.E. Kalman, Algebraic geometric description of the class of linear systems of constant dimension, in *8th Annual Princeton Conference on Information Sciences and Systems*, vol. 3 (Princeton, N.J., 1974)

R.E. Kalman, P. Falb, M.A. Arbib, *Topics in Mathematical System Theory*. International Series in Pure and Applied Mathematics. (McGraw Hill, 1969)

A. Lindquist, G. Picci, *Linear Stochastic Systems; A Geometric Approach to Modeling*. Series in Contemporary Mathematics, vol. 1 (Springer, Berlin [u.a.], 2015). ISBN 978-3-662-45749-8; 3-662-45749-0

Models with Exogenous Variables

8

In many applications the observed variables can be divided into variables which are to be explained by a model, called endogenous variables or outputs, and observed variables which influence the endogenous variables without being influenced by them. The latter variables are called exogenous variables or observed inputs. They provide important information about the endogenous variables. This applies for instance in control applications, where we want to steer the "outputs" by "inputs" which are under our control. Another wide range of application is concerned with prediction, where we try to enhance the prediction of the endogenous variables by the use of information provided by the exogenous variables.

8.1 General Structure

Let (x_t) and (y_t) denote the processes of the exogenous and endogenous variables, respectively. We assume that the endogenous variables are generated as

$$y_t = \hat{y}_t + u_t, \quad \text{where } \hat{y}_t = \sum_{j=0}^{\infty} l_j x_{t-j} \tag{8.1}$$

i.e. the endogenous variables y_t are modeled as the sum of a linear function of the past and present values of the exogenous variables x_t and, as the explanation by the exogenous variables is not exact in general, a "noise term" u_t. Hence, one may interpret y_t as the noisy outputs of a linear system with inputs x_t. In order to simplify the discussion we assume that the stacked process $(w_t = (x_t', u_t')')$ is stationary and centered ($\mathbf{E}w_t = 0$) and that $\underline{l}(L) = \sum_{j \geq 0} l_j L^j$ is an l_1 filter (i.e. the coefficients are absolutely summable). Note that we require, that $\underline{l}$ is a *causal* filter. This is a reasonable assumption, if $\hat{y}_t$ is interpreted as the output of a real-world physical (technical) system with inputs x_t. Of course, causality is also particularly relevant,

© Springer Nature Switzerland AG 2022

M. Deistler and W. Scherrer, *Time Series Models*, Lecture Notes in Statistics 224,

https://doi.org/10.1007/978-3-031-13213-1_8

when the goal is prediction. In addition, we assume that the noise process (u_t) is *orthogonal* to the input process (x_t), i.e.

$$\mathbf{E}u_t x_s' = 0 \ \forall t, s \in \mathbb{Z}. \tag{8.2}$$

This assumption is, e.g. justified, if u_t represents measurement errors of the outputs, where the measurement errors are independent from the inputs of the system under consideration. Furthermore (8.2) implies that $\hat{y}_t$ is the best approximation of y_t by a linear function of past, present *and* future exogenous variables x_s. Compare also the discussion concerning the Wiener filter in Sect. 4.4. However, since we in addition assume that l is causal, we here assume that future values are not needed for the best approximation.

If (x_t) and (u_t) have a spectral density then, in an obvious notation, we obtain

$$f_y(\lambda) = l(\lambda) f_x(\lambda) l(\lambda)^* + f_u(\lambda) \tag{8.3}$$
$$f_{yx}(\lambda) = l(\lambda) f_x(\lambda). \tag{8.4}$$

Now let us assume that the noise (u_t) has a rational spectral density and that the transfer function of the filter $l(L)$ is rational. Then, by proposition 6.7 and the corollary 6.8, we get

$$(y_t) = \underline{l}(L)(x_t) + \underline{k}(L)(\epsilon_t).$$

Here (ϵ_t) is the innovation process of (u_t) with

$$\mathbf{E}\epsilon_t \epsilon_t' = \Sigma \text{ and } \mathbf{E}\epsilon_t x_s' = 0 \, \forall t, s \in \mathbb{Z}.$$

The two rational filters $\underline{l}, \underline{k}$ are stable, $\underline{k}$ is minimum phase and $\underline{k}(0) = I_n$. Note that under the persistence of excitation condition

$$f_x(\lambda) > 0 \ \mu \text{ a.e.} \tag{8.5}$$

and under the condition $\Sigma > 0$, the filters $\underline{l}, \ \underline{k}$ and the noise variance Σ are uniquely determined from the joint spectrum of the observed processes $(x_t', y_t')'$.

Proposition 8.1 *Under the above assumptions ($\underline{k}, \underline{l}$ are stable, rational filters, $\underline{k}$ is minimum phase, $\underline{k}(0) = I_n$, $\Sigma > 0$ and under the persistence of excitation condition (8.5)) the filters $\underline{k}, \underline{l}$ and the innovation variance Σ are uniquely determined from the joint spectrum of the observed process $(x_t', y_t')'$.*

Proof From Eq. (8.4), we get

$$l(\lambda) = f_{yx}(\lambda) f_x^{-1}(\lambda)$$

and from (8.3) it follows that

$$f_y(\lambda) - f_{yx}(\lambda) f_x^{-1}(\lambda) f_{yx}(\lambda)^* = f_u(\lambda) = k(\lambda) \Sigma k(\lambda)^*$$

which by proposition 6.7 implies the uniqueness of $\underline{k}$ and Σ. $\square$

In the following two sections, we will drop the constraint $\underline{k}(0) = I_n$. In this slightly more general setting, the innovation process of the noise process (u_t) is $(\underline{k}(0)\epsilon_t)$ and $(\underline{k}, \underline{l}, \Sigma)$ and $(\tilde{\underline{k}}, \tilde{\underline{l}}, \tilde{\Sigma})$ are observationally equivalent (i.e. they generate the same spectral densities f_{yx} and f_y for given f_x) if and only if there exists a non-singular matrix $T \in \mathbb{R}^{n \times n}$ such that

$$\tilde{\underline{l}} = \underline{l}, \ \tilde{\underline{k}} = \underline{k}T^{-1} \text{ and } \tilde{\Sigma} = T\Sigma T'. \tag{8.6}$$

If we assume that we know future exogenous variables, then prediction of the $y_t's$ is straightforward. For example, the one-step ahead prediction for y_{t+1}, given the infinite past of $(x_s', y_s')'$, $s \le t$ and the future value x_{t+1} is equal to

$$\hat{y}_{t,1} = \sum_{j \ge 0} l_j x_{t+1-j} + \sum_{j > 0} k_j \epsilon_{t+1-j}$$

and the corresponding prediction error is ϵ_{t+1}. This is also called "conditional prediction", since it is based on the assumption that the future value x_{t+1} is known. In the case that the exogenous variables act on the outputs with a certain delay (revival time), i.e. where $l_0 = 0$ holds, the above prediction is feasible even if the future value x_{t+1} is not known.

A number of model classes arise from the general form (8.1) by choosing a particular parametrization of the filters $\underline{k}$ and $\underline{l}$.

ARMAX   model (AutoRegressive Moving Average model with eXogenous variables): Set $\underline{l}(z) = \underline{a}^{-1}(z)\underline{d}(z)$ and $\underline{k}(z) = \underline{a}^{-1}(z)\underline{b}(z)$ with suitably chosen polynomial matrices $\underline{a}, \underline{b}, \underline{d}$. This model may be written as a difference equation:

$$\underline{a}(L)(y_t) = \underline{d}(L)(x_t) + \underline{b}(L)(\epsilon_t)$$

ARX   model: If we set $\underline{b}(z) = I_n$ then as a special case the ARX model is obtained

$$\underline{a}(L)(y_t) = \underline{d}(L)(x_t) + (\epsilon_t)$$

which may also be written as a regression model

$$y_t = d_0 x_t + d_1 x_{t-1} + \cdots + d_r x_{t-r} + a_1 y_{t-1} + \cdots + a_p y_{t-p} + \epsilon_t.$$

Note that all variables on the right-hand side, except for ϵ_t, are observed (as opposed to the ARMAX case) and hence the model parameters may be estimated by ordinary least squares. It is easy to derive equations of Yule–Walker type in this case and then to show that the parameters of the model are uniquely determined from the autocovariance function of the stacked process $(x_t', y_t')'$, if Σ is positive definite and if the following persistence of excitation condition holds:

$$\mathbb{E}x_t^{r+1}(x_t^{r+1})' > 0$$

Furthermore, prediction is particularly easy for ARX models. The conditional one-step ahead prediction for y_{t+1} is

$$\hat{y}_{t,1} = d_0 x_{t+1} + d_1 x_t + \cdots + d_r x_{t+1-r} + a_1 y_t + \cdots + a_p y_{t+1-p}$$

OE model (Output Error model): $\underline{k}(z) = I_n$, i.e. the outputs here are corrupted by *white* noise $y_t = \hat{y}_t + u_t$, $(u_t = \epsilon_t) \sim WN(\Sigma)$. This model class is often used in system and control applications.

SSX State Space models with eXogenous inputs: A particular choice for such models is

$$s_{t+1} = A s_t + E x_t + B \epsilon_t \tag{8.7}$$
$$y_t = C s_t + F x_t + \epsilon_t \tag{8.8}$$

which corresponds to $\underline{k}(z) = C(z^{-1} I_m - A)^{-1} B + I_n$ and $\underline{l}(z) = C(z^{-1} I_m - A)^{-1} E + F$.

For prediction in this case, see the previous Sect. 7.4 concerning the Kalman filter.

The discussions concerning identifiability of ARMA models (in Sect. 6.3) and state space models (in Sect. 7.3) also apply here. As we have shown in Proposition 8.1 the filters $\underline{k}, \underline{l}$ are identified under general assumptions, however, e.g. the parameters $\underline{a}$, $\underline{b}$, $\underline{d}$ of the ARMAX models are only identified, when we impose additional identifying restrictions. This topic is dealt with in the next two sections.

8.2 Structural Identifiability of ARX and ARMAX Models

In many cases, multivariate ARX or ARMAX systems (also called VARX and VARMAX, respectively, systems) resulting from particular applications contain additional a-priori restrictions coming from (e.g. physical or economic) theory. In particular, often a-priori information is available, that certain variables do not influence other variables, resulting in so-called zero restrictions on parameters. A detailed discussion of the parametrization of ARMAX systems with such a-priori restriction is given in Deistler (1983).

We consider ARMAX models of the form

$$a_0 y_t = a_1 y_{t-1} + \cdots + a_p y_{t-p} + d_0 x_t + \cdots + d_r x_{t-r} + \epsilon_t + b_1 \epsilon_{t-1} + \cdots + b_q \epsilon_{t-q} \tag{8.9}$$

where $d_j \in \mathbb{R}^{n \times m}$, $a_j, b_j \in \mathbb{R}^{n \times n}$ and $(\epsilon_t) \sim WN(\Sigma)$ is white noise. Note that, in this section, we drop the assumption $a_0 = I_n$. On the other hand, we assume that there are a number of a-priori restrictions on the parameters, coming from an underlying theory. In particular, we will consider the case that certain entries of the parameter matrices are either zero or equal to one. The model and the model parameters have a theory-based interpretation and hence the above form (including the a priori restrictions on the parameters) is called structural form.

Example (structural ARX system)

A small and simple macroeconomic system is given in Schönfeld (1971). This is a structural ARX system of the form:

$$
\begin{aligned}
\text{private consumption} \quad C_t &= \alpha_0 + \alpha_1 Y_t + \alpha_2 C_{t-1} + \epsilon_{1t} \\
\text{private net-investments} \quad I_t &= \kappa(C_t - C_{t-1}) + \epsilon_{2t} \\
\text{imports} \quad Im_t &= \delta_0 + \delta_1 Y_t + \delta_2 Im_{t-1} + \epsilon_{3t} \\
\text{net-income} \quad Y_t &= C_t + I_t + G_t + Ex_t - Im_t
\end{aligned}
$$

Here the endogenous variables are C_t, I_t, Im_t and Y_t and the exogenous variables are exports Ex_t, government expenditures G_t and the constant $Z_t \equiv 1$. The parameters $\alpha_0, \alpha_1, \alpha_2, \kappa, \delta_0, \delta_1, \delta_2$ are structural parameters.

Note that the fourth equation is a so-called definition equation with known coefficients and no error term. As a consequence, the assumption $\Sigma > 0$ is no longer valid. However, we will not discuss this case $(\det(\Sigma) = 0)$ in detail.

Let $\underline{a}(z) = a_0 - a_1 z - \cdots - a_p z^p$ and $\underline{b}(z) = I_n + b_1 z + \cdots + b_q z^q$. Throughout we assume stability, i.e.

$$
\det \underline{a}(z) \neq 0 \ |z| \leq 1 \tag{8.10}
$$

as well as the minimum-phase condition

$$
\det \underline{b}(z) \neq 0 \ |z| < 1.
$$

Note that (8.10) implies $\det a_0 = \det \underline{a}(0) \neq 0$. Furthermore, we require that the exogenous variables (x_t) and the noise (ϵ_t) are orthogonal to each other

$$
\mathbf{E} x_s \epsilon_t' = 0 \ \forall s, t \in \mathbb{Z}
$$

and that (x_t) has a spectral density f_x which satisfies the persistence of excitation condition (8.5):

$$
f_x(\lambda) > 0 \ \mu \ \text{a.e.} \tag{8.11}
$$

In addition, throughout we assume

$$
\Sigma = \mathbf{E} \epsilon_t \epsilon_t' > 0.
$$

First let us consider the ARX case $(q = 0)$

$$
a_0 y_t = a_1 y_{t-1} + \cdots + a_p y_{t-p} + d_0 x_t + \cdots + d_r x_{t-r} + \epsilon_t \tag{8.12}
$$

and the corresponding unique stationary solution:

$$(y_t) = \underline{l}(L)(x_t) + \underline{k}(L)(\epsilon_t) \text{ where } \underline{l} = \underline{a}^{-1}\underline{d}, \ \underline{k} = \underline{a}^{-1}. \tag{8.13}$$

If we premultiply (8.12) by a_0^{-1}, we obtain the so called *reduced form*

$$y_t = a_0^{-1}a_1 y_{t-1} + \cdots + a_0^{-1}a_p y_{t-p} + a_0^{-1}d_0 x_t + \cdots + a_0^{-1}d_r x_{t-r} + a_0^{-1}\epsilon_t. \tag{8.14}$$

Clearly the representation (8.13) satisfies all assumptions of the previous section and hence we may use Proposition 8.1 (and the notes thereafter) to discuss the identifiability of the structural form parameters:

Proposition 8.2 *Two ARX systems $(\underline{a}(z), \underline{d}(z), \Sigma)$ and $(\tilde{\underline{a}}(z), \tilde{\underline{d}}(z), \tilde{\Sigma})$, satisfying the above assumptions, are observationally equivalent if and only if there exists a constant non-singular matrix T such that*

$$(\tilde{\underline{a}}(z), \tilde{\underline{d}}(z)) = T(\underline{a}(z), \underline{d}(z)) \tag{8.15}$$

$$\tilde{\Sigma} = T\Sigma T' \tag{8.16}$$

hold.

Proof First note, that we here consider the general case, where $\underline{k}(0)$ is not restricted to $\underline{k}(0) = I_n$. Thus, we immediately get from (8.6) that $\tilde{\Sigma} = T\Sigma T'$. By the identity $\tilde{\underline{k}} = \underline{k}T^{-1}$, it follows that $\tilde{\underline{a}}^{-1} = \tilde{\underline{k}} = \underline{k}T^{-1} = \underline{a}^{-1}T^{-1}$ and thus $\tilde{\underline{a}} = T\underline{a}$. Finally by $\tilde{\underline{l}} = \underline{l}$, we get $\tilde{\underline{a}}^{-1}\tilde{\underline{d}} = \tilde{\underline{l}} = \underline{l} = \underline{a}^{-1}\underline{d}$ and thus $\tilde{\underline{d}} = \tilde{\underline{a}}\underline{a}^{-1}\underline{d} = T\underline{d}$.   □

Now write (8.15) as

$$(\tilde{a}_0, \tilde{a}_1, \ldots, \tilde{a}_p, \tilde{d}_0, \tilde{d}_1, \ldots, \tilde{d}_r) = T(a_0, a_1, \ldots, a_p, d_0, d_1, \ldots, d_r). \tag{8.17}$$

Clearly a given parameter space is identifiable, if it contains additional a-priori restrictions (for all elements in this parameter space), such that T in (8.17) is necessarily the identity matrix. The following set of conditions has been proposed in Koopmans et al. (1950):

$$\text{Every row of } (a_0, a_1, \ldots, a_p, d_0, d_1, \ldots, d_r)$$
$$\text{contains at least } n - 1 \text{ (a-priori) zero restrictions.} \tag{8.18}$$

Let A_i denote the submatrix consisting of those columns of $(a_0, \ldots, a_p, d_0, \ldots, d_r)$ where zero restrictions in the i-th row occur. Then we assume

$$\text{rk}(A_i) = n - 1 \text{ for } i = 1, \ldots, n. \tag{8.19}$$

Now (8.17) implies

$$(u_i T) A_i = 0$$

where $u_i \in \mathbb{R}^{1 \times n}$ is the i-th (row) unit vector in $\mathbb{R}^{1 \times n}$ and thus, by (8.19) T must be diagonal. Assuming in addition

$$\text{the diagonal elements of } a_0 \text{ are equal to one} \qquad (8.20)$$

we conclude, that under these restrictions, we have identifiability.

Proposition 8.3 *Under the assumptions of the above Proposition 8.2 and if the parameter space in addition satisfies (8.18), (8.19) and (8.20) then the corresponding class of ARX systems is identifiable.*

Exercise 8.4 Prove that the condition $a_0 = I_n$ satisfies the requirements (8.18), (8.19) and (8.20). This shows again that the reduced form is identifiable.

In order to give a general picture, we introduce the parameter space $\Theta_{p,r} = \{(a_0, \ldots, a_p, d_0, \ldots, d_r, \Sigma) \mid (\underline{a}(z)\} \text{ is stable and } \Sigma > 0\}$. Because of Proposition 8.2 the parameter space consists of equivalence classes of parameters which correspond, for given spectral density f_x, to the same spectral densities f_{yx} and f_y. Each equivalence class contains a unique element, which corresponds to the reduced form.

In case of identifiability, along the lines of Proposition 8.3, one may distinguish two cases:

Exact identifiability: This is the situation, where in each of the above equivalence classes a unique element is selected by the restrictions.

Over identifiablility: This is a situation, where some of the equivalence classes are discarded by the restrictions. Thus over identifiability means that the zero restrictions discard transfer functions (and thus input- output behavior), whereas this is not the case for exact identifiability. Over identifiability also imposes restrictions on the reduced form parameters, thus these parameters are not "free", which makes estimation more demanding.

Finally, we consider the case, where in a VARMAX system

$$\underline{a}(\mathrm{L})(y_t) = \underline{d}(\mathrm{L})(x_t) + \underline{b}(\mathrm{L})(\epsilon_t),$$

a number of polynomial elements in $(\underline{a}, \underline{d})$ are known to be zero, i.e. we know that certain outputs or inputs do *not* occur in certain equations at any time lag. In most cases one does not have structural information about the noise and hence here we only assume that $\underline{b}(0) = b_0 = I_n$. Assuming persistence of excitation (i.e. (8.5)), it follows from the notes after Proposition 8.1, that two systems $(\underline{a}, \underline{b}, \underline{d})$ and $(\tilde{\underline{a}}, \tilde{\underline{b}}, \tilde{\underline{d}})$ are observationally equivalent if and only

$$\tilde{\underline{a}}^{-1}\tilde{\underline{d}} = \underline{a}^{-1}\underline{d} \text{ and } \tilde{\underline{a}}^{-1}\tilde{\underline{b}} = \underline{a}^{-1}\underline{b}T^{-1}.$$

Clearly this condition is equivalent to the existence of a (rational) matrix $\underline{u}$ and of a non-singular matrix T such that

$$(\tilde{\underline{a}}, \tilde{\underline{b}}, \tilde{\underline{d}}) = \underline{u}(\underline{a}, \underline{b}T^{-1}, \underline{d}). \tag{8.21}$$

From this observation, we then obtain (see Hannan (1971))

Proposition 8.5 (Structural identifiability in ARMAX classes) *Consider a class of ARMAX systems, where the general assumptions mentioned above and the following conditions hold:*

(1) In every row of $(\underline{a}, \underline{d})$, there are at least $n - 1$ elements prescribed to be zero.
(2) The polynomial submatrix A_i of $(\underline{a}, \underline{d})$ consisting of those columns, where the prescribed zeros in the i-th row occur, has rank $n - 1$ (a.e.).
(3) Every row of $(\underline{a}, \underline{b}, \underline{d})$ is relatively prime.
(4) The diagonal elements of $\underline{a}(0)$ are equal to one.

Then the class is identifiable.

Proof From (1) we have

$$u_i A_i = 0$$

where u_i denotes the i-th row of $\underline{u}$ in (8.21). By condition (2) then all elements in u_i have to be zero, except for the i-th one; thus $\underline{u}$ is a diagonal matrix; by (3) $\underline{u}(z)$ must be constant and by (4) the identity matrix. Finally, the condition $\underline{b}(0) = I_n$ implies $T = I_n$ and thus the class is identifiable.   □

Of course, there are many other possibilities, for imposing structural information on VARMAX system. For example, one may also consider restrictions on the entries of the (individual) parameter matrices a_i and/or d_i. See also Deistler (1978).

8.3   Structural (V)AR Models

In the last decades, *structural* (vector) autoregressive (SVAR) models have become very popular in macroeconomics, finance and related fields. Here we only give a very short introduction to this model class. For an excellent overview see, e.g. the book by Kilian and Lütkepohl (2017), which in particular also contains a number of concrete examples of SVAR models in macroeconomics. Although this model does not contain exogenous variable, we include it here in this chapter, since the discussion about structural identifiability is closely related to the discussion of the previous sections.

As has been noted above, SVAR models are often used for analysis of macroeconomic data from the background of economic theory. For the discussion of these models, we assume that the n-dimensional observed process (x_t) collects some macroeconomic variables, e.g. gross domestic product, inflation rate, etc. We start with the stationary case, i.e. with an AR system (5.1), which satisfies the stability condition. Therefore, (ϵ_t) is the innovation process, which means that the ϵ_t's represent the *unanticipated changes* in the considered economic variables. In addition, we assume that the innovation variance $\Sigma = \mathbf{E}\epsilon_t\epsilon_t' > 0$ is positive definite. The key idea of SVAR models is to construct "structural shocks", v_{kt}, $k = 1, \ldots, n$ say, from the innovations, which have an "economic interpretation", e.g. as "demand shocks" or "technology shocks". The structural shocks represent unanticipated changes in the observed economic variables and they are assumed to be mutually uncorrelated. Often one also assumes that they are standardized to unit variance. Hence, we define the vector of structural shocks $v_t = (v_{1t}, \ldots, v_{nt})'$ by

$$A\epsilon_t = Bv_t$$

and require

$$I_n = \mathbf{E}v_t v_t' = B^{-1}A\Sigma A'(B')^{-1}. \tag{8.22}$$

This means that the structural shocks are orthogonalized innovations. If we multiply the AR system (5.1) from the left by A and plug in the vector of shocks v_t then we obtain the so-called "AB-model":

$$Ax_t = \tilde{a}_1 x_{t-1} + \cdots + \tilde{a}_p x_{t-p} + Bv_t \tag{8.23}$$

with $\tilde{a}_j = Aa_j$, for $j = 1, \ldots, p$. The above representation is the "structural form", and (5.1) is called the "reduced form" of the AR system. Both forms are equivalent representations of the AR process (x_t). The parameters of the reduced form (5.1) are uniquely determined from the autocovariance function of the process (x_t) (see, e.g. the discussion on the Yule–Walker equations in Sect. 5.4), whereas the structural form parameters are not unique without additional conditions. The process (x_t) has a causal MA(∞) representation, see (5.5),

$$x_t = \sum_{j \geq 0} k_j \epsilon_{t-j} = \sum_{j \geq 0} \tilde{k}_j v_{t-j} \text{ with } \tilde{k}_j = k_j(A^{-1}B).$$

The sequence of coefficients $(\tilde{k}_j \mid j \geq 0)$ is the (orthogonalized) *impulse response* of the SVAR model. The coefficient $\tilde{k}_j$ represents the dependence of the future value x_{t+j} on the current shocks v_t. In particular, $\tilde{k}_0 = A^{-1}B$ represents the immediate response of the observed variables to the shocks. Here we have used $k_0 = I_n$, see (5.6).

Exercise 8.6 (forecast error variance decomposition) The above MA(∞) representation with the structural shocks is also the basis for the so-called "Forecast Error Variance Decomposition (FEVD)". Show that the variance of the h-step ahead prediction error (from the infinite past) for the k-th variable ($x_{t+h,k}$) is equal to

$$\Sigma_{h,kk} = \sum_{j=0}^{h-1}\sum_{i=1}^{n} \tilde{k}_{j,ki}^2 = \sum_{i=1}^{n}\sum_{j=0}^{h-1} \tilde{k}_{j,ki}^2$$

where $\Sigma_{h,kk}$ denotes the k-th diagonal element of the matrix Σ_h and $\tilde{k}_{j,ki}$ is the (k,i)-th entry of the matrix $\tilde{k}_j$. Therefore, the ratio

$$\left(\sum_{j=0}^{h-1}\tilde{k}_{j,kl}^2\right) \Big/ \left(\sum_{i=1}^{n}\sum_{j=0}^{h-1}\tilde{k}_{j,ki}^2\right)$$

is a measure for the contribution of the l-th structural shocks to the future of the k-th variable.

Of course the condition $I_n = \mathbf{E}v_t v_t'$ is not sufficient to uniquely determine the matrices A, $B \in \mathbb{R}^{n\times n}$. A simple heuristic argument (counting free parameters and independent equations) shows that one needs $n^2 + n(n-1)/2$ additional conditions. Note that A, B contain $2n^2$ parameters and that (8.22) gives $n(n+1)/2$ independent equations. These additional conditions should be based on economic reasoning, in order to give the shocks $v_t = B^{-1}A\epsilon_t$ an economic meaning. In the econometric literature, a number of identifying restrictions have been proposed. The reason for using two matrices A and B is that this gives more flexibility for imposing restrictions. Often one sets $B = I_n$ and only uses the A matrix with suitable restrictions. Alternatively, one may set $A = I_n$ or one may use both matrices.

The identifying restrictions in many cases impose a specified structure to the matrix $\tilde{k}_0$. In particular, one often assumes that some of the entries $\tilde{k}_0$ are equal to zero. If the (i,j)-th entry of this matrix is zero, then the j-th structural shock does *not* have a contemporaneous impact on the i-th observed variable. Conditions of this type are hence called "short-run restrictions". The most common identification scheme postulates that (after possibly reordering of the variables) the matrix $\tilde{k}_0$ is lower triangular, which means that the last $(n-j+1)$ variables respond instantaneously to the j-th component of the shock, whereas the first $(j-1)$ variables only react with a delay. Correspondingly, we may assume that A is lower triangular with ones on the diagonal and that B is a diagonal matrix with non-negative diagonal entries. Any stable AR model with a non-singular innovation variance may be rewritten as an AB-model of this form and, under these restrictions, the matrices A, B and $\tilde{a}_j$, $j = 1,\ldots,p$ are uniquely determined. This follows immediately by noting that $\Sigma = A^{-1}(BB')(A^{-1})'$ is the Cholesky decomposition of Σ.

If the economic variables under consideration are integrated of order one, then often an AR model for the first differences is used, i.e. we model $x_t = (z_t - z_{t-1}) =$

Δz_t, where z_t contains the variables of interest or the corresponding log-values[1]. The effect of the shocks v_t at time t on the future levels (or log-values) $z_{t+j} = z_{t-1} + x_t + \cdots + x_{t+j}$ is the sum of the effects on the (future) differences, i.e. is equal to $\sum_{j=0}^t k_j \epsilon_t = \sum_{j=0}^t \tilde{k}_j v_t$. Therefore the matrix

$$\tilde{\Theta} := \sum_{j \geq 0} \tilde{k}_j = \Theta(A^{-1}B), \quad \text{where } \Theta = \left(\sum_{j \geq 0} k_j\right)$$

is a measure for the "long-run effect" of the structural shocks. A typical assumption then is that a structural shock has *no* long run effect on some specified variables, i.e. the respective entries of $\tilde{\Theta}$ are assumed to be zero.

Of course one may also mix "levels" and first differences, i.e. x_t may be of the form

$$x_t = \begin{pmatrix} z_{1t} \\ z_{2t} - z_{2,t-1} \end{pmatrix}$$

and correspondingly one may combine short-run and long-run restrictions.

Another option to deal with stationary and integrated variables, is to consider cointegrated processes, as discussed in Sect. 5.5. In particular, let us assume that $(x_t = z_t)$ is a cointegrated process which has an AR representation as discussed in the "Granger representation theorem". Then the observed process has the representation (5.22), i.e.

$$x_t = \Xi \sum_{j=1}^t \epsilon_j + v_t + x_0^* = \tilde{\Xi} \sum_{j=1}^t v_j + v_t + x_0^*$$

where

$$\Xi := \underline{k}(1) \quad \text{and} \quad \tilde{\Xi} = \Xi(A^{-1}B).$$

The entries of the matrix $\tilde{\Xi}$ represent the long-run effects of the structural shocks on (the levels of) the considered variables and hence are called long-run multiplier. In order to get identifying restrictions for the matrices A, B we may impose restrictions (with an analogous interpretation as above) on $\tilde{\Xi}$. However, note that this matrix is singular, if the process is cointegrated. In particular, note that the rows of Ξ, which correspond to stationary components of (x_t), contain only zeros. Hence, e.g. imposing a zero restriction on an entry in such a row, does not help to identify A, B.

As has been stated at the very beginning of this section, here we only give a very short introduction to structural VAR models. Therefore many questions are not touched, e.g.

- In general, it is a very challenging task to find identifying restrictions, which are on the one hand economic reasonable and which are on the other hand easy to handle.

---

[1] In this case the first differences $x_t = (z_t - z_{t-1}) = \Delta z_t$ are approximate growth rates.

- As has been stated above, $n^2 + n(n-1)/2$ restrictions for identifiability of A, B are needed. Of course, in general, a mere counting of restrictions is not sufficient. For example, for linear restrictions, one has to prove that these restrictions are linearly independent (see also the rank condition in Proposition 8.3) and for non-linear restrictions, things get even more complicated. Compare also the above remark on restrictions on the (singular) matrix $\tilde{\Xi}$.
- If the structural form parameters are *just identified*, i.e. if the restrictions are sufficient to uniquely determine the parameters but do not exclude any AR models, then the estimation is usually straightforward. Typically, one starts with an estimate of the reduced form model and then converts this reduced form to the structural form. On the other hand, if the structural form parameters are *over identified*, i.e. if only a subset of AR models may be cast as an SVAR models of the required form, then estimation is much more involved.

References

M. Deistler, The structural identifiability of linear models with autocorrelated errors in the case of cross-equation restrictions. J. Econom. **8**(1), 23–31 (1978)

M. Deistler, The Properties of the parameterization of ARMAX systems and their relevance for structural estimation and dynamic specification. Econometrica **51**(4), 1187–1207 (1983). ISSN 00129682, 14680262

E.J. Hannan, The identification problem for multiple equation systems with moving average errors. Econometrica **39**(5), 751–765 (1971). ISSN 00129682, 14680262. http://www.jstor.org/stable/1909577

L. Kilian, H. Lütkepohl, *Structural Vector Autoregressive Analysis* (Cambridge University Press, Cambridge, 2017). 1316647331

T.C. Koopmans, H. Rubin, R.B. Leipnik, Measuring the equation systems of dynamic economics, in *Cowles Commission Monograph*, vol. 10, ed. by T.C. Koopmans (Wiley, New York, 1950), pp.53–237

P. Schönfeld, *Methoden der Ökonometrie; Band II* (Verlag Franz Vahlen GmbH, 1971)

Granger Causality

9

In many areas, the dependence structure between variables is of special interest. In particular, often the question arises whether there exist *causal* relationships. The topic of this chapter is the discussion of the so-called Granger causality, which has been introduced in Granger (1969). This concept has received great attention in econometrics—but also in many other fields—since it allows to empirically test for causal relations given time series data. Discussions and characterizations of Granger causality may be found, e.g. in Sims (1972), Pierce and Haugh (1977), Geweke (1984).

9.1 Granger Causality

We consider a regular, stationary process which is partitioned into three sub-processes (x_t', y_t', z_t'), where the sub-processes (x_t), (y_t), (z_t) have dimensions n_1, n_2 and n_3 respectively. In the following discussion the sub-process (z_t) may also be missing. We consider the linear, least squares, one-step ahead prediction for x_{t+1} given different "information sets", e.g. we consider the one-step ahead prediction for x_{t+1} given the own past $\mathbb{H}_t(x)$ or the prediction given the past of all sub-processes $\mathbb{H}_t(x, y, z) = \overline{sp}\{x_s, y_s, z_s \mid s \leq t\}$. Quite analogously the Hilbert spaces $\mathbb{H}_t(x, y)$ and $\mathbb{H}_t(x, z)$ are defined.

Definition 9.1 The (sub) process (y_t) is *(Granger) causal* for (x_t), if

$$P_{\mathbb{H}_t(x)} x_{t+1} \neq P_{\mathbb{H}_t(x,y)} x_{t+1}$$

i.e. if the covariance matrices of the respective one-step ahead prediction errors are not equal:

$$\mathbf{Var}(x_{t+1} - P_{\mathbb{H}_t(x)} x_{t+1}) \neq \mathbf{Var}(x_{t+1} - P_{\mathbb{H}_t(x,y)} x_{t+1}).$$

Note that clearly $\mathbf{Var}(x_{t+1} - P_{\mathbb{H}_t(x)} x_{t+1}) \geq \mathbf{Var}(x_{t+1} - P_{\mathbb{H}_t(x,y)} x_{t+1})$ holds.

© Springer Nature Switzerland AG 2022
M. Deistler and W. Scherrer, *Time Series Models*, Lecture Notes in Statistics 224,
https://doi.org/10.1007/978-3-031-13213-1_9

If

$$P_{\mathbb{H}_t(x)} x_{t+1} \neq P_{\mathbb{H}_t(x,y) \oplus \mathrm{sp}\{y_{t+1}\}} x_{t+1}$$

then we say that (y_t) is instantaneously causal for (x_t).

If (y_t) is causal for (x_t) and vice versa (x_t) is causal for (y_t), then we say that there is a feedback between (x_t) and (y_t).

Granger causality means that the prediction for x_{t+1} is improved if we add the information about the past of (y_t). The ordering in time is the characteristic feature of this definition. The process (y_t) is Granger causal for (x_t) if the past of (y_t) is influential for (x_t), in the sense that the *prediction* of x_{t+1} is improved by the knowledge of the past of (y_t). However, this concept may lead to false conclusions due to the "post hoc ergo propter hoc" fallacy: An event in the future is not necessarily caused by an event in the past. For a more detailed discussion of this point see, e.g. Kilian and Lütkepohl (2017, Chap. 7).

The stationarity of the processes considered makes this definition independent of the time point $(t + 1)$ considered. Clearly for singular processes this definition does not make sense. It should also be noted that non-linear effects are not taken into account by the definition above.

The next, more general, definition takes the additional information contained in the past of the process (z_t) into account.

Definition 9.2 If

$$P_{\mathbb{H}_t(x,z)} x_{t+1} \neq P_{\mathbb{H}_t(x,y,z)} x_{t+1}$$

then (y_t) is, conditional on (z_t), causal for (x_t).

The following proposition gives a characterization of Granger causality:

Proposition 9.3 *Let $(x'_t, y'_t)'$ be a stationary process with a rational spectral density*

$$f(\lambda) = \begin{pmatrix} f_x(\lambda) & f_{xy}(\lambda) \\ f_{yx}(\lambda) & f_y(\lambda) \end{pmatrix}$$

where $f(\lambda) > 0$, $\lambda \in [-\pi, \pi]$. Furthermore, the Wold representation of the process (using an obvious notation) is

$$\begin{pmatrix} x_t \\ y_t \end{pmatrix} = \underline{k}(L)(\epsilon_t) = \begin{pmatrix} \underline{k}_x(L) & \underline{k}_{xy}(L) \\ \underline{k}_{yx}(L) & \underline{k}_y(L) \end{pmatrix} \begin{pmatrix} \epsilon^x_t \\ \epsilon^y_t \end{pmatrix}, \quad \text{with } \mathbf{E} \begin{pmatrix} \epsilon^x_t \\ \epsilon^y_t \end{pmatrix} \begin{pmatrix} \epsilon^x_t \\ \epsilon^y_t \end{pmatrix}' = \begin{pmatrix} \Sigma_x & \Sigma_{xy} \\ \Sigma_{yx} & \Sigma_y \end{pmatrix}.$$

Note that $\underline{k}(z)$ is the stable and strictly minimum-phase spectral factor of $\underline{f}(z)$ which has been discussed in Proposition 6.7. This spectral factor satisfies the normalization condition $\underline{k}(0) = I$. The AR(∞) representation of $(x'_t, y'_t)'$ is denoted by

$$\begin{pmatrix} x_t \\ y_t \end{pmatrix} = \sum_{j \geq 1} \begin{pmatrix} a^{xx}_j & a^{xy}_j \\ a^{yx}_j & a^{yy}_j \end{pmatrix} \begin{pmatrix} x_{t-j} \\ y_{t-j} \end{pmatrix} + \begin{pmatrix} \epsilon^x_t \\ \epsilon^y_t \end{pmatrix} = \tilde{\underline{a}}(L) \begin{pmatrix} x_t \\ y_t \end{pmatrix} + \begin{pmatrix} \epsilon^x_t \\ \epsilon^y_t \end{pmatrix}.$$

The following statements are equivalent:

(1) (y_t) is not Granger causal for (x_t).
(2) The AR(∞) representation is block lower triangular, i.e.

$$a_j^{xy} = 0, \quad j = 1, 2, 3, \dots. \tag{9.1}$$

(3) The spectral factor $\underline{k}(z)$ is block lower triangular, i.e.

$$\underline{k}_{xy}(z) = 0. \tag{9.2}$$

(4) The transfer function of the Wiener filter

$$\underline{l}(z) = \underline{f}_{yx}(z)\underline{f}_x^{-1}(z)$$

has no poles within or on the unit circle and therefore the transfer function has a representation as a power series which converges within and on the unit circle, i.e. the Wiener filter is causal.

Proof Note that the one-step ahead prediction for $(x'_{t+1}, y'_{t+1})'$ from the past $\mathbb{H}_t(x, y)$ is given by

$$\begin{pmatrix} \hat{x}_{t,1} \\ \hat{y}_{t,1} \end{pmatrix} = \sum_{j \geq 1} \begin{pmatrix} a_j^{xx} & a_j^{xy} \\ a_j^{yx} & a_j^{yy} \end{pmatrix} \begin{pmatrix} x_{t+1-j} \\ y_{t+1-j} \end{pmatrix}$$

and thus the equivalence of (1) and (2) is evident. The spectral factor $\underline{k}(z)$ and the AR(∞) representation are connected by

$$(I - \tilde{\underline{a}}(z)) = \underline{k}^{-1}(z)$$

and therefore (2)$\Longleftrightarrow$ (3) is evident from the inversion formula of block triangular matrices.

(4) $\Longrightarrow$ (1): If the Wiener Filter is causal, then the projection of y_{ks} onto the time domain $\mathbb{H}(x)$ is contained in the past $\mathbb{H}_s(x)$ and thus for $s \leq t$ we have $P_{\mathbb{H}(x)} y_{ks} = P_{\mathbb{H}_t(x)} y_{ks}$. The Hilbert space $\mathbb{H}_t(x, y) = \overline{\mathrm{sp}}\{x_s, y_s \mid s \leq t\}$ therefore is the orthogonal sum of $\mathbb{H}_t(x)$ and a space which is orthogonal to $\mathbb{H}(x)$. This implies that the projection of $x_{k,t+1}$ onto $\mathbb{H}_t(x, y)$ is contained in $\mathbb{H}_t(x)$ which means that (y_t) is not Granger causal for (x_t).

(3) $\Longrightarrow$ (4): The condition (3) implies

$$\underline{l}(z) = \underline{f}_{yx}(z)\underline{f}_x^{-1}(z) = (\underline{k}_{yx}(z)\Sigma_x + \underline{k}_y(z)\Sigma_{yx})\underline{k}_x^*(z) \left(\underline{k}_x(z)\Sigma_x\underline{k}_x^*(z)\right)^{-1}$$

$$= (\underline{k}_{yx}(z)\Sigma_x + \underline{k}_y(z)\Sigma_{yx})\Sigma_x^{-1}\underline{k}_x^{-1}(z)$$

The spectral factor $\underline{k}(z)$ has no poles for $|z| \leq 1$ and clearly this also holds for all submatrices. The spectral factor has no zeros for $|z| \leq 1$, in other words the inverse transfer function $\underline{k}(z)$ has no poles on and inside the unit circle. Due to the block triangular structure this also holds for the inverse of the left upper block, i.e. $\underline{k}_x^{-1}(z)$ has no poles for $|z| \leq 1$. This concludes the proof.   □

As has been noted above, the transfer function $\underline{k}(z)$ corresponds to the Wold representation

$$\begin{pmatrix} x_t \\ y_t \end{pmatrix} = \sum_{j \geq 0} \begin{pmatrix} k_j^{xx} & k_j^{xy} \\ k_j^{yx} & k_j^{yy} \end{pmatrix} \begin{pmatrix} \epsilon_{t-j}^x \\ \epsilon_{t-j}^y \end{pmatrix}$$

of the joint process $(x_t', y_t')'$. Clearly, the transfer function is block lower triangular if and only if $k_j^{xy} = 0$ for all $j \geq 0$. In this case $x_t = \sum_{j \geq 0} k_j^{xx} \epsilon_{t-j}^x$ is the Wold representation of the sub-process (x_t) and the h-step ahead prediction of x_{t+h} given the own past $\mathbb{H}_t(x)$ and the h-step ahead prediction given $\mathbb{H}_t(x, y)$ are identical

$$P_{\mathbb{H}_t(x,y)} x_{t+h} = \sum_{j \geq h} k_j^{xx} \epsilon_{t+h-j}^x + \underbrace{k_j^{xy} \epsilon_{t+h-j}^y}_{=0} = P_{\mathbb{H}_t(x)} x_{t+h}.$$

Thus, we have shown that

$$\left(P_{\mathbb{H}_t(x,y)} x_{t+h} = P_{\mathbb{H}_t(x)} x_{t+h} \, \forall h > 0 \right) \text{ if and only if } \left(P_{\mathbb{H}_t(x,y)} x_{t+1} = P_{\mathbb{H}_t(x)} x_{t+1} \right).$$

This observation explains why in the above definition for Granger causality only the one-step ahead predictions are used. If the one-step ahead prediction is not improved by considering the past of (y_t) then the same is true for all h-step ahead predictions. However, as will be detailed below, the situation changes if we consider conditional Granger causality.

The above proposition is the basis for statistical tests for non-causality. Non-causality may, e.g. be tested by testing the null-hypothesis $a_j^{xy} = 0$, $j = 1, 2, \ldots$.

Using analogous arguments, one may prove the following proposition:

Proposition 9.4 *Under the assumptions of the above proposition, the following statements are equivalent:*

(1) (y_t) *is not instantaneously causal for* (x_t).
(2) *The transfer function* $\underline{k}(z)$ *is block lower triangular as in (9.2) and the covariance matrix of the innovations is block diagonal:*

$$\mathbf{E} \begin{pmatrix} \epsilon_t^x \\ \epsilon_t^y \end{pmatrix} \begin{pmatrix} \epsilon_t^x \\ \epsilon_t^y \end{pmatrix}' = \begin{pmatrix} \Sigma_x & 0 \\ 0 & \Sigma_y \end{pmatrix}. \tag{9.3}$$

(3) *The AR(∞) representation satisfies (9.1) and the innovation covariance satisfies (9.3).*

(4) *The transfer function* $\underline{l}(z) = \underline{f}_{yx}(z)\underline{f}_x^{-1}(z)$ *of the Wiener filter is causal and satisfies* $\underline{l}(0) = 0$.

If (y_t) is not instantaneously causal for (x_t), then we may use (x_t) as an exogenous variable in a model for (y_t) as has been discussed in the Sect. (8.1).

The following proposition deals with conditional causality:

Proposition 9.5 *Let* (x_t', y_t', z_t') *be stationary with rational spectral density* f, *where* $f(\lambda) > 0$, $\lambda \in [-\pi, \pi]$. *Then* (y_t) *is, conditional on* (z_t), *not causal for* (x_t) *if and only if the AR(∞) representation of the process (in obvious notation)*

$$\begin{pmatrix} x_t \\ y_t \\ z_t \end{pmatrix} = \begin{pmatrix} \tilde{\underline{a}}^{xx}(L) & \tilde{\underline{a}}^{xy}(L) & \tilde{\underline{a}}^{xz}(L) \\ \tilde{\underline{a}}^{yx}(L) & \tilde{\underline{a}}^{yy}(L) & \tilde{\underline{a}}^{yz}(L) \\ \tilde{\underline{a}}^{zx}(L) & \tilde{\underline{a}}^{zy}(L) & \tilde{\underline{a}}^{zz}(L) \end{pmatrix} \begin{pmatrix} x_t \\ y_t \\ z_t \end{pmatrix} + \begin{pmatrix} \epsilon_t^x \\ \epsilon_t^y \\ \epsilon_t^z \end{pmatrix}$$

satisfies

$$\tilde{\underline{a}}^{xy}(L) = 0. \tag{9.4}$$

The analysis of conditional causality is more complicated than the analysis of (unconditional) causality. First note that the answer, of course, depends on the additional information contained in the process (z_t). Depending on the process (z_t), the process (y_t) may or may not be (conditionally) causal for (x_t). It is intuitively clear that it is "harder" to see a causal relation if the information in (z_t) gets richer. For example, there may be a causal relation from (y_t) to (z_t) and from (z_t) to (x_t). If we ignore (z_t), then a "spurious" causal relation between (y_t) and (x_t) may occur, which leads to a misinterpretation of the dependence structure between these variables. Secondly, we remark that the h-step ahead prediction for some $h > 1$ may be improved by taking the past of (y_t) into account, although the 1-step ahead prediction is *not* improved. Therefore in the literature sometimes an alternative definition (see, e.g. Lütkepohl 2005) for conditional (non-) causality is used: (y_t) is, conditionally on (z_t), Granger non-causal for (x_t) if

$$P_{\mathbb{H}_t(x,y,z)} x_{t+h} = P_{\mathbb{H}_t(x,z)} x_{t+h} \; \forall h \geq 1.$$

One practical problem with this definition is that it imposes non linear restrictions on the AR(∞) coefficients and hence is much harder to test. See Proposition 9.5 and the Exercise 9.10, dealing with the AR(1) case, below.

Exercise 9.6 Prove the Proposition 9.4.

Exercise 9.7 Prove the Proposition 9.5.

Exercise 9.8 One may also define "instantaneous influence" as follows: The process (y_t) is instantaneously influential for (x_t) if

$$P_{\mathbb{H}_t(x,y)} \, x_{t+1} \neq P_{\mathbb{H}_t(x,y)\oplus \text{sp}\{y_{t+1}\}} \, x_{t+1}.$$

Prove that (y_t) is *not* instantaneously influential for (x_t) if and only if the covariance matrix of the innovations of the joint process $(x_t', y_t')'$ is block diagonal, i.e. if the condition (9.3) is satisfied. This shows that "instantaneous influence" is a symmetric relation between processes.

Exercise 9.9 A conditional version is defined as follows: The process (y_t) is, conditionally on (z_t), instantaneously influential for (x_t) if

$$P_{\mathbb{H}_t(x,y,z)\oplus \text{sp}\{z_{t+1}\}} \, x_{t+1} \neq P_{\mathbb{H}_t(x,y,z)\oplus \text{sp}\{y_{t+1},z_{t+1}\}} \, x_{t+1}.$$

Let

$$S = \begin{pmatrix} S_x & S_{xy} & S_{xz} \\ S_{yx} & S_y & S_{yz} \\ S_{zx} & S_{zy} & S_z \end{pmatrix} = \Sigma^{-1}$$

denote the inverse of the covariance matrix Σ of the innovations (ϵ_t) of the joint process $(x_t', y_t', z_t')'$. Prove that (y_t) is *not* conditionally, instantaneously influential for (x_t) if and only if the block $S_{xy} = 0 = S_{yx}'$ is zero. This shows that also "conditional, instantaneous influence" is a symmetric relation between processes.

Exercise 9.10 Suppose that the joint process (x_t', y_t', z_t') is an AR(1) process defined by

$$\begin{pmatrix} x_t \\ y_t \\ z_t \end{pmatrix} = \begin{pmatrix} a_{11} & a_{12} & a_{13} \\ a_{21} & a_{22} & a_{23} \\ a_{31} & a_{32} & a_{33} \end{pmatrix} \begin{pmatrix} x_{t-1} \\ y_{t-1} \\ z_{t-1} \end{pmatrix} + \begin{pmatrix} \epsilon_t^x \\ \epsilon_t^y \\ \epsilon_t^z \end{pmatrix}.$$

Prove that $P_{\mathbb{H}_t(x,y,z)} \, x_{t+1} = P_{\mathbb{H}_t(x,z)} \, x_{t+1}$ is equivalent to $a_{12} = 0$. Now consider the two-step ahead prediction. Derive an analogous condition for

$$P_{\mathbb{H}_t(x,y,z)} \, x_{t+2} = P_{\mathbb{H}_t(x,z)} \, x_{t+2}.$$

This methodology is particularly easy to handle for the case of multivariate AR models. In the AR case, it is straightforward to test for Granger non-causality, and it is also easy to develop estimates which take a particular dependence structure into account.

The causal interdependence may also be discussed via graphs, see, e.g. Eichler (2006). The vertices of such a causality graph represent the variables (i.e. the one-dimensional component processes) and the edges show the causal interdependencies. Often one considers mixed graphs, i.e. graphs which contain different types of edges, corresponding to different types of dependencies. For example, $x \rightarrow y$ may represent (conditional) Granger causality of (y_t) on (x_t) and x—y may represent (conditional) instantaneous influence. These (and related) graphs provide a method for the visualization and analysis of the dependency structure of high-dimensional times series, respectively processes.

Exercise 9.11 As a simple example consider the trivariate $(n = 3)$ AR(1) process

$$\begin{pmatrix} x_t \\ y_t \\ z_t \end{pmatrix} = \begin{pmatrix} a_{11} \, a_{12} \, a_{13} \\ a_{21} \, a_{22} \, a_{23} \\ a_{31} \, a_{32} \, a_{33} \end{pmatrix} \begin{pmatrix} x_{t-1} \\ y_{t-1} \\ z_{t-1} \end{pmatrix} + \begin{pmatrix} \epsilon_t^x \\ \epsilon_t^y \\ \epsilon_t^z \end{pmatrix}, \quad S = \begin{pmatrix} s_x \, s_{xy} \, s_{xz} \\ s_{yx} \, s_y \, s_{yz} \\ s_{zx} \, s_{zy} \, s_z \end{pmatrix} = \Sigma^{-1}$$

and the following graph.

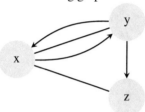

Here the edges represent conditional dependencies. For example, $y \rightarrow z$ means that (y_t) is, conditionally on (x_t), Granger causal for (z_t) and z—x means that (z_t) is, conditionally on (y_t), instantaneously influential for (x_t) (and vice versa). Which zero restrictions does this graph impose on the parameters a_{ij} and s_{ij}?

References

M. Eichler, Graphical modelling of dynamic relationships in multivariate time series, in *Handbook of Time Series Analysis*. ed. by B.S.J.T.M. Winterhalder (Wiley-VCH, Berlin, 2006)

J. Geweke, Inference and causality in economic time series models, in *Handbook of Econometrics*, vol. 2, Chapter 19 (Elsevier, 1984), pp. 1101–1144. https://doi.org/10.1016/S1573-4412(84)02011-0. http://www.sciencedirect.com/science/article/pii/S1573441284020110

C. Granger, Investigating causal relations by econometric models and cross-spectral methods. Econometrica **37**(3), 424–438 (1969). ISSN 00129682, 14680262. http://www.jstor.org/stable/1912791

L. Kilian, H. Lütkepohl, *Structural Vector Autoregressive Analysis* (Cambridge University Press, Cambridge, 2017). 1316647331

H. Lütkepohl, *New Introduction to Multiple Time Series Analysis* (Springer, Berlin, 2005)

D.A. Pierce, L.D. Haugh, Causality in temporal systems. J. Economet. **5**(3), 265–293 (1977). ISSN 0304-4076. https://doi.org/10.1016/0304-4076(77)90039-2. http://www.sciencedirect.com/science/article/pii/0304407677900392

C.A. Sims, Money, income, and causality. Am. Econ. Rev. **62**(4), 540–552 (1972). ISSN 00028282. http://www.jstor.org/stable/1806097

Dynamic Factor Models

10

In this chapter we deal with linear dynamic factor models and related topics, such as dynamic principal component analysis (dynamic PCA). A main motivation for the use of such models is the so-called "curse of dimensionality" plagueing modeling of high dimensional time series by "ordinary" multivariate AR or ARMA models: For instance, consider an AR system (5.1) for a, say, 20-dimensional time series. Then each of the matrices a_i contains 400 "free" parameters, if no additional restrictions on the parameter space have been imposed, i.e. in such a case the parameter spaces grow with n^2, whereas the data, for given sample size T, grow with n. Thus for moderate T and large n (as is the case, e.g. in many situations faced in macroeconomics), reliable parameter estimation in "fully parametrized" AR(X) or ARMA(X) models is hardly possible. On the other hand, e.g. in macroeconomics, for analysis and in particular for short-term forecasting, modeling of "comovement" and of "cross-sectional dependencies" between a large number of univariate time series recently has received increasing attention and appropriate tools for modeling of high-dimensional time series have been developed. Correspondingly, during the last 25 years, a substantial literature has emerged, dealing with such models, methods and applications, in particular for factor models in this context.

For the case of i.i.d observations, (static) factor models have a long history dating back to Spearman (1904) and Burt (1909). For the dynamic case, these models have been developed much later, see, e.g. Sargent and Sims (1977), Geweke (1977), Scherrer and Deistler (1998), Forni et al. (2000), Stock and Watson (2002), Bai and Ng (2002), Bai (2003), Doz et al. (2011), Hallin et al. (2020) and Lippi et al. (2022). The aim of this chapter is to give an introduction to dynamic factor models (in the broad sense), rather than to treat such models in detail.

© Springer Nature Switzerland AG 2022
M. Deistler and W. Scherrer, *Time Series Models*, Lecture Notes in Statistics 224,
https://doi.org/10.1007/978-3-031-13213-1_10

10.1   Dynamic Factor Models—General Structure

In general terms, we here consider models of the form

$$(x_t) = \underline{L}(\mathrm{L})(f_t) + (u_t),\tag{10.1}$$

where (x_t) is the n-dimensional, observed process, $(z_t) = \underline{L}(\mathrm{L})(f_t)$ is the so-called process of *latent variables*, (f_t) is the r-dimensional ($r < n$) process of *factors* and (u_t) is *noise*. The linear dynamic, $(n \times r)$-dimensional, filter $\underline{L}$ is called the *factor loading matrix*. In most cases it is assumed that $\underline{L}$ is an l_1-filter. We assume that (z_t), (f_t) and (u_t) are not (directly) observed and that (f_t) (and thus (z_t)) and (u_t) are stationary, centered and that their spectral densities exist. In addition, we require that the factors and the noise are orthogonal to each other, i.e.

$$\mathbf{E}f_t u_s' = 0 \ \forall t, s.\tag{10.2}$$

Note that, in this chapter, we use the symbol Σ (instead of f) for spectral densities in order to avoid possible confusion with the factors f_t. Then, using an obvious notation, (10.1) and (10.2) imply the following relation between the spectral densities of (x_t), (f_t) and (u_t) and the factor loading $\underline{L}$

$$\Sigma_x(\lambda) = L(\lambda)\Sigma_f(\lambda)L(\lambda)^* + \Sigma_u(\lambda).\tag{10.3}$$

In many cases, we assume that the $(n \times r)$-dimensional filter $\underline{L}$ is rational, i.e. $\underline{L}(z)$ is a rational matrix, and that $\underline{L}(z)$ has full rank r (a.e.). Due to $r < n$ and $\mathrm{rk}(\underline{L}(z)) = r$ (a.e.), the rational matrix $\underline{L}(z)$ has a non-trivial, $n - r$ dimensional left kernel, consisting of n-dimensional (row) vectors with rational entries. Let $\underline{c}(z)$ denote a matrix, whose rows form a basis for the left kernel of $\underline{L}(z)$; as easily can be seen, $\underline{c}(z)$ can be chosen as a polynomial matrix. Then

$$\underline{c}(z)\underline{L}(z) = 0$$

implies

$$c_0 z_t + c_1 z_{t-1} + \cdots + c_h z_{t-h} = 0, \ c_i \in \mathbb{R}^{(n-r)\times n},$$

i.e. there is an exact (i.e. noise free) linear dynamic relation between the components of (z_t). This gives a precise meaning to the notion of "comovement" between the component processes in (z_t).

Two common interpretations for factor models are as follows: In the first one, the latent variables (z_t) may be interpreted as the (true) unobserved inputs and outputs of a linear dynamic deterministic system and (u_t) are measurement errors. Note that in this context we don't need to make an a priori classification into inputs and outputs and both may be corrupted by noise. This corresponds to an "errors-in-variables" model, see, e.g. Scherrer and Deistler (1998). An alternative interpretation of (10.1) is the decomposition of the observations into a part (z_t) representing comovement and a second part (u_t) representing "idiosyncratic", i.e. "component specific", noise. An example for this would be to disentangle a vector of stock market returns into market caused and firm specific returns.

A central task for modeling of factor models is to uniquely disentangle the latent variables (z_t) from the noise (u_t), given the observed variables (x_t) or their second moments. Furthermore one has to (uniquely) determine the factors (f_t) and the factor loading matrix $\underline{L}$ for given latent variables $(z_t) = \underline{L}(L)(f_t)$. For doing this, additional assumptions are needed, where different assumptions lead to different classes of factor models:

- Idiosyncratic noise: Here it is assumed that the noise components are mutually orthogonal $\mathbf{E}u_{it}u_{js} = 0$ for $i \neq j$ and $t, s \in \mathbb{Z}$, which is equivalent to the condition that the spectral density Σ_u is diagonal. Therefore, the linear relations between the observed variables are exclusively due to the common factors.
  In the case, where (f_t) and (u_t) are white noise and where the factor loading matrix $\underline{L}(L) = L$ is constant, we obtain the classical factor analysis model, see, e.g. Lawley and Maxwell (1971). In a (general) dynamic factor model with strictly idiosyncratic noise, the loading matrix as well as the factor and noise processes may be dynamic. This great flexibility comes at a certain price, the question of identifiability and hence also estimation is much more involved, than in the classical factor analysis model, see Scherrer and Deistler (1998).
- Weakly idiosyncratic noise: Here the above assumption of idiosyncratic noise is weakened, such that "weak" cross-sectional dependency is allowed for the noise components. On the other hand, it is assumed that the latent variables are "strongly" coupled. These concepts are defined in a setting where the cross-sectional dimension n grows to infinity. The corresponding models are called generalized static or dynamic, respectively, factor models and will be discussed in Sect. 10.3.
- A further option to disentangle factors and noise is to assume that the noise process (u_t) is a white noise process. Here the dynamics, i.e. the serial dependencies, of the observed variables is entirely attributed to the loading matrix and the factor variables. See, e.g. Diversi et al. (2007).
- The cointegration model discussed in Sect. 5.5 may be interpreted as a factor model, however, in a non-stationary setting. The Beveridge–Nelson decomposition (5.20) is a factor model (with a constant factor loading matrix $\underline{L} = L \in \mathbb{R}^{n \times r}$) where the common trend components (= factors) are integrated, whereas the remainder (= noise) is stationary.
- In the next chapter, we will shortly present the factor GARCH model (see (11.6)). This is a model, where the factor variables are (scalar) GARCH processes (i.e. conditionally heteroscedastic processes), whereas the remainder (= noise) has a constant conditional variance.
- In many applications, the aim is simply to approximate the observed variables by a linear combination of $r < n$ factor variables, i.e. the task is to construct the factors and the factor loading matrix such that the approximation error (= noise $(u_t) = (x_t) - \underline{L}(L)(f_t))$ is as small as possible. Typically, the approximation is

rated with the mean squared error $\mathbf{E}u_t' u_t \longrightarrow$ min. In the case where the factor loading matrix is assumed to be constant, this leads to static principal component analysis, whereas the general case leads to the dynamic principal component analysis, which will be discussed in the next Sect. 10.2.

The above list is by no means complete and we will only discuss some special cases in the following.

We now consider the problem of determining the factor loadings and the factors from given latent variables. Clearly the "factorization" $(z_t) = \underline{L}(L)(f_t)$ is by no means unique for given (z_t). For example, we may squeeze in a non-singular matrix $T \in \mathbb{R}^{r \times r}$ such that $(z_t) = (\underline{L}(L)T)(T^{-1}f_t)$. Note that even the number r of factors is not determined by (z_t).

If the latent variables have a rational spectral density $\underline{\Sigma}_z$, with rank q (a.e.), then there exists a factorization, corresponding to the Wold decomposition,

$$\underline{\Sigma}_z(z) = \underline{L}(z)\underline{L}(z)^*$$

where $\underline{L}(z)$ is a rational $n \times q$ matrix, which is stable and minimum-phase (i.e. $\mathrm{rk}(\underline{L}(z)) = q$ for all $|z| < 1$). This follows from a generalization of Proposition 6.7 to the case of a rank deficient spectral density. Furthermore, there exists a q-dimensional white noise process $(\epsilon_t) \sim \mathrm{WN}(I_q)$ such that

$$(z_t) = \underline{L}(L)(\epsilon_t) \tag{10.4}$$

where $(\underline{L}(0)\epsilon_t)$ is the innovation process of (z_t). As easily can be seen, the (dynamic) factors ϵ_t are unique up to pre-multiplication by orthogonal matrices and q is the *minimum* number of factors for given $\underline{\Sigma}_z$.

In some special, but important, cases it is possible to find a representation with a constant factor loading $\underline{L}(L) = L \in \mathbb{R}^{n \times r}$, i.e. the "dynamics" of the latent variables is completely attributed to the factors. As a simple example consider a model

$$z_t = L_0 \epsilon_t + L_1 \epsilon_{t-1} = \underbrace{(L_0, L_1)}_{=:K} \begin{pmatrix} \epsilon_t \\ \epsilon_{t-1} \end{pmatrix} = K f_t.$$

Here we get a "static" factor loading matrix $K = (L_0, L_1)$ and a so-called "static" factor $f_t = (\epsilon_t', \epsilon_{t-1}')'$. A general approach to construct a static factor loading matrix, if the latent variables (z_t) have a rational spectral density, is as follows. We start with the Wold decomposition (10.4) and construct a state space representation as detailed in Sect. 7.3:

$$s_{t+1} = As_t + B\epsilon_t \tag{10.5}$$
$$z_t = Cs_t + D\epsilon_t = (C, D)(s_t', \epsilon_t')'. \tag{10.6}$$

If the state dimension, m say, is small enough, i.e. if $m + q < n$, then we have a representation with a static matrix $K = (C, D)$ and corresponding factors $f_t =$

$(s_t', \epsilon_t')'$. Another approach uses the covariance matrix $\gamma_z(0) = \mathbf{E}z_t z_t'$ of the latent variables. If this matrix has rank $r < n$ then there exists a factorization as $\gamma_z(0) = LL'$ with $L \in \mathbb{R}^{n \times r}$ and $\mathrm{rk}(L) = r$. Furthermore it follows that $z_t = Lf_t$, where $f_t = (L'L)^{-1}L'z_t$, i.e. we have a representation with a static matrix $L \in \mathbb{R}^{n \times r}$ and corresponding factors.

We call a factor model *static*, if $\underline{L}(\mathrm{L})$ in (10.1) is a constant matrix $L \in \mathbb{R}^{n \times r}$. In this case, f_t are called *static factors*. If $\underline{L}(\mathrm{L})$ is not a constant, we speak of a *dynamic* factor model and *dynamic factors* f_t.

As the above discussion shows, it is often, but not always, possible to construct a static model for a given process of latent variables. Note that, for given (z_t), the minimal number, q say, of components of a dynamic factor is equal to the rank of the spectral density $\underline{\Sigma}_z$. As the above discussion shows, the minimal number, r say, of components of a static factor is equal to the rank of the covariance matrix $\gamma_z(0) = \mathbf{E}z_t z_t'$ of z_t. Note in addition, that $q = \mathrm{rk}(\underline{\Sigma}_z) \leq r = \mathrm{rk}(\gamma_z(0)) \leq n$ holds.

A commonly used factor model is of the form

$$x_t = Lf_t + u_t \tag{10.7}$$

where $L \in \mathbb{R}^{n \times r}$ is a constant factor loading matrix of rank $r < n$ and (f_t) is an autoregressive process of the form

$$f_t = a_1 f_{t-1} + \cdots a_p f_{t-p} + b\epsilon_t \tag{10.8}$$

where $(\epsilon_t) \sim \mathrm{WN}(I_q)$ is q-dimension white noise, $b \in \mathbb{R}^{r \times q}$ has full rank $q \leq r$, and

$$\underline{a}(z) = I_r - a_1 z - \ldots - a_p z^p$$

satisfies the stability condition. Now, we assume that the noise components (u_{it}) are e.g. autoregressive of order one

$$u_{it} = \rho_i u_{i,t-1} + v_{it}, \ |\rho_i| < 1, \ i = 1, \ldots, n \tag{10.9}$$

and that the white noise processes $(v_{it}) \sim \mathrm{WN}(\sigma_i^2)$ are mutually orthogonal to each other ($\mathbf{E}v_{it}v_{js} = 0$ for $i \neq j$). Then we may cast this factor model (10.7)–(10.9) into the following state space form (see, e.g. Doz et al. (2011)):

$$s_{t+1} = \begin{pmatrix} a_1 & \cdots & a_{p-1} & a_p & 0 \\ I_q & & 0 & 0 & 0 \\ & \ddots & & & \\ 0 & & I_q & 0 & 0 \\ 0 & \cdots & 0 & 0 & \rho \end{pmatrix} s_t + \begin{pmatrix} b\epsilon_{t+1} \\ 0 \\ \vdots \\ 0 \\ v_{t+1} \end{pmatrix}$$

$$x_t = \begin{pmatrix} L & 0 & \cdots & 0 & I_n \end{pmatrix} s_t$$

where

$$s_t = (f_t', f_{t-1}' \ldots, f_{t+1-p}', u_t')', \ v_t = (v_{1t}, \ldots, v_{nt})' \text{ and } \rho = \mathrm{diag}(\rho_1, \ldots, \rho_n).$$

An advantage of this model structure is, that it is relatively easy to estimate the model parameters and, for given model parameters, the factors and the latent variables may be estimated from the observed variables. For both tasks, the Kalman filter and the Kalman smoother are valuable tools. The model structure is relatively simple and hence the model is parsimonious, in particular, as compared to full AR or ARMA models. Furthermore, this model structure (and generalizations) often yields reasonable results, e.g. in prediction for high-dimensional time series. A survey on state space modeling of factor models has been given in Poncela et al. (2021).

10.2   Principal Components in the Frequency Domain

We assume that the reader is familiar with "ordinary" principal component analysis (PCA) commencing from the covariance matrix of an n dimensional random vector x. For principal components in the frequency domain, or dynamic PCA (DPCA), the decomposition on the right-hand side of (10.1) is obtained by solving the minimization problem: Minimize

$$\mathbf{E}u_t'u_t = \operatorname{tr}\mathbf{E}u_tu_t' = (2\pi)^{-1}\int_{-\pi}^{\pi}\operatorname{tr}\Sigma_u(\lambda)d\lambda$$

for $(u_t) = (x_t) - \underline{L}(L)(f_t)$ and a given number of r factors. This gives an (in the mean squares sense) optimal approximation of the observed variables in terms of r factors and a corresponding factor loading matrix. Principal component analysis in the frequency domain has been developed by Brillinger (2001, Chap. 9).

The "static" PCA simply restricts the loading matrix to be a static transformation $\underline{L}(L) = L \in \mathbb{R}^{n \times r}$. The static PCA is a very powerful and popular method in multivariate statistics and is used as a workhorse in many different scenarios. The computations are based on the eigenvalue decomposition of the variance $\mathbf{E}x_tx_t'$ of the observed variables.

It is remarkable that the dynamic PCA mimics this approach in the sense that it is based on a "frequency-wise" eigenvalue decomposition of the spectral density of the process (x_t). In order to present the basics, we repeat some of the results of Chaps. 3 and 4 and introduce some more notation. The frequency domain of a stationary process (x_t) is the Hilbert space $\mathbb{H}_F(x) = \mathbb{L}_2^{\mathbb{C}}([-\pi, \pi], \mathcal{B}, F)$ of (equivalence classes) of functions $a: [-\pi, \pi] \longrightarrow \mathbb{C}^{1 \times n}$ which are square integrable with respect to the spectral distribution function F of this stationary process. In this and the next section, we will only deal with processes which have a spectral density, Σ_x say. Correspondingly we will use the shorthand notation $\mathbb{H}_F(x) = \mathbb{L}_2^{\mathbb{C}}(\Sigma_x)$. The frequency domain is isometrically isomorphic to the (complex) time domain $\mathbb{H}^{\mathbb{C}}(x)$ of the process. The stochastic integral

$$\Phi_x^{-1}(a) = \int_{-\pi}^{\pi} a(\lambda)dz(\lambda) \in \mathbb{H}^{\mathbb{C}}(x)$$

defines an isometry, which maps functions $a \in \mathbb{H}_F(x)$ to elements in the time domain of the process. We can also construct a process, $(y_t = U^t \, \Phi_x^{-1}(a) \,|\, t \in \mathbb{Z})$ say, for a given $a \in \mathbb{H}_F(x)$. Here U is the forward shift in the time domain $\mathbb{H}^{\mathbb{C}}(x)$, see Proposition 1.15. Therefore, we interpret $a \in \mathbb{H}_F(x)$ as the transfer function of a linear filter. With a slight abuse of notation this filter is denoted with $\underline{a}(\mathrm{L})$, i.e. $(y_t = U^t \, \Phi_x^{-1}(a) \,|\, t \in \mathbb{Z}) = \underline{a}(\mathrm{L})(x_t)$. This identification of elements of the frequency domain with transfer functions of linear, dynamic filters is especially simple, if a has a Fourier series expansion $a(\lambda) = \sum_{-\infty}^{\infty} a_j e^{-i\lambda j}$ with absolutely summable coefficients $\sum_j \|a_j\| < \infty$. In this case the corresponding filter is the l_1-filter, given by $\underline{a}(\mathrm{L}) = \sum_{j=-\infty}^{\infty} a_j \, \mathrm{L}^j$. Correspondingly, matrix functions $a \in (\mathbb{H}_F(x))^m$ can be interpreted as the transfer function of a linear filter, which maps the n-dimensional process (x_t) to an m-dimensional process $(y_t) = \underline{a}(\mathrm{L})(x_t)$.

Now, let us return to the above approximation problem. We only give a sketch of the solution and refer to Brillinger (2001) and Forni and Lippi (2001) for more details. Initially, we allow non-zero correlations between the noise and the latent variables and hence consider a decomposition of the spectral density Σ_x as

$$\Sigma_x = \Sigma_z + \Sigma_{zu} + \Sigma_{uz} + \Sigma_u.$$

The only requirement for this decomposition is that

$$\begin{pmatrix} \Sigma_z & \Sigma_{zu} \\ \Sigma_{uz} & \Sigma_u \end{pmatrix}$$

is a spectral density and that $\Sigma_z(\lambda)$ has rank less than or equal to r (μ-a.e. on $[-\pi, \pi]$). For a fixed, but arbitrary frequency λ, the optimal decomposition of $\Sigma_x(\lambda)$ may be computed as follows. Consider the eigenvalue decomposition of $\Sigma_x(\lambda)$:

$$\Sigma_x(\lambda) = \sum_{k=1}^{n} o_k(\lambda) \rho_k(\lambda) o_k^*(\lambda)$$

where $o_k(\lambda) \in \mathbb{C}^{n \times 1}$ are the normalized eigenvectors and $\rho_1(\lambda) \geq \rho_2(\lambda) \geq \cdots \geq \rho_n(\lambda) \geq 0$ are the associated eigenvalues. Note that $o_k^*(\lambda) o_j(\lambda) = 0$ for $k \neq j$ and $o_k^*(\lambda) o_k(\lambda) = 1$. Then $\Sigma_z^{\mathrm{PC}}(\lambda) = \sum_{k=1}^{r} o_k(\lambda) \rho_k(\lambda) o_k^*(\lambda)$, $\Sigma_{zu}^{\mathrm{PC}}(\lambda) = 0$ and $\Sigma_u^{\mathrm{PC}}(\lambda) = \sum_{k=r+1}^{n} o_k(\lambda) \rho_k(\lambda) o_k^*(\lambda)$ is the optimal decomposition for this frequency, since

$$\mathrm{tr} \, \Sigma_u^{\mathrm{PC}}(\lambda) \leq \mathrm{tr} \, \Sigma_u(\lambda)$$

holds for any decomposition of the spectral density Σ_x, which satisfies the above requirements.

It remains to show that this frequency-wise decomposition of the spectral density corresponds to a decomposition of the observed process (x_t). To this end, we assume that the spectral density $\Sigma_x(\lambda)$ is a continuous function of the frequency λ and that $\Sigma_x(\lambda)$ has distinct eigenvalues $\rho_1(\lambda) > \rho_2(\lambda) > \cdots > \rho_n(\lambda) \geq 0$ for all frequencies. (The results may also be proved for the case of multiple eigenvalues, however with even more technical difficulties.) Then the following statements hold (see Forni and Lippi (2001)):

(1) The eigenvalues $\rho_k \colon [-\pi, \pi] \to \mathbb{R}$, are Lebesgue-measurable functions and integrable in $[-\pi, \pi]$. Note that $\rho_k(\lambda)$ is a continuous function of the matrix $\Sigma_x(\lambda)$ and thus ρ_k (as a continuous function of a continuous function) is a continuous function of the frequency λ. The function ρ_k is integrable due to

$$\int_{-\pi}^{\pi} \rho_k(\lambda)d\lambda \le \int_{-\pi}^{\pi} \sum_{l=1}^{n} \rho_l(\lambda)d\lambda = \int_{-\pi}^{\pi} \mathrm{tr}(\Sigma_x(\lambda))d\lambda < \infty.$$

(2) It can be shown that the normalized eigenvectors $o_k \colon [-\pi, \pi] \to \mathbb{C}^n$ are Lebesgue-measurable, bounded and square integrable with respect to the measure Σ_x on $[-\pi, \pi]$. In particular, integrability follows from

$$\int_{-\pi}^{\pi} o_k^*(\lambda)\Sigma_x(\lambda)o_k(\lambda)d\lambda = \int_{-\pi}^{\pi} \rho_k(\lambda)d\lambda < \infty \text{ and}$$

$$\mathrm{tr}\left[\int_{-\pi}^{\pi} o_k(\lambda)o_k^*(\lambda)\Sigma_x(\lambda)o_k(\lambda)o_k^*(\lambda)d\lambda\right] = \int_{-\pi}^{\pi} \rho_k(\lambda)d\lambda < \infty.$$

In other words $o_k^* \in \mathbb{L}_2^{\mathbb{C}}(\Sigma_x)$ and $o_k o_k^* \in (\mathbb{L}_2^{\mathbb{C}}(\Sigma_x))^n$. Therefore

$$(f_{kt}^{PC}) = o_k^*(\mathrm{L})(x_t), \quad (f_t^{PC}) = ((f_{1t}^{PC}, \ldots, f_{rt}^{PC})')$$

$$(z_t^{PC}) = \sum_{k=1}^{r} o_k(\mathrm{L})((f_{kt}^{PC})) = \left(\sum_{k=1}^{r} o_k o_k^*(\mathrm{L})\right)(x_t)$$

$$(u_t^{PC}) = (x_t) - (z_t) = \left(\sum_{k=r+1}^{n} o_k o_k^*(\mathrm{L})\right)(x_t)$$

are well-defined stationary processes. Note also that due to the symmetry properties of the spectral density ($\Sigma_x(\lambda) = \Sigma_x(-\lambda)'$) these processes are real valued.

(3) The processes (f_t^{PC}), (z_t^{PC}), (u_t^{PC}) are jointly stationary with spectral densities:

$$\Sigma_f^{PC}(\lambda) = \begin{pmatrix} o_1^*(\lambda) \\ \vdots \\ o_r^*(\lambda) \end{pmatrix} \Sigma_x(\lambda) \left(o_1(\lambda) \cdots o_r(\lambda)\right) = \mathrm{diag}(\rho_1(\lambda), \ldots, \rho_r(\lambda))$$

$$\Sigma_z^{PC}(\lambda) = \left(\sum_{k=1}^{r} o_k(\lambda)o_k^*(\lambda)\right) \Sigma_x(\lambda) \left(\sum_{k=1}^{r} o_k(\lambda)o_k^*(\lambda)\right) = \sum_{k=1}^{r} o_k(\lambda)\rho_k(\lambda)o_k^*(\lambda)$$

$$\Sigma_u^{PC}(\lambda) = \left(\sum_{r+1}^{n} o_k(\lambda)o_k^*(\lambda)\right) \Sigma_x(\lambda) \left(\sum_{r+1}^{n} o_k(\lambda)o_k^*(\lambda)\right) = \sum_{r+1}^{n} o_k(\lambda)\rho_k(\lambda)o_k^*(\lambda)$$

$$\Sigma_{uf}^{PC}(\lambda) = \left(\sum_{k=r+1}^{n} o_k(\lambda) o_k^*(\lambda) \right) \Sigma_x(\lambda) \big(o_1(\lambda) \cdots o_r(\lambda) \big) = 0 \in \mathbb{C}^{n \times r}$$

$$\Sigma_{uz}^{PC}(\lambda) = \left(\sum_{k=r+1}^{n} o_k(\lambda) o_k^*(\lambda) \right) \Sigma_x(\lambda) \left(\sum_{k=1}^{r} o_k(\lambda) o_k^*(\lambda) \right) = 0 \in \mathbb{C}^{n \times n}.$$

The results are summarized in the following proposition:

Proposition 10.1 *The dynamic principal component analysis model, as defined above, minimizes* $\mathbf{E} u_t' u_t$ *among all factor models with at most r factors. The dynamic PCA model is observable (in the sense that the factors, latent variables and the errors are linear transformations of the observed process) and orthogonal (i.e. the factors and the noise are orthogonal to each other).*

The factors f_{kt} are called the dynamic principal components associated with the process (x_t).

The mean squared error of the DPCA models is

$$\mathbf{E}\left((u_t^{PC})' u_t^{PC} \right) = \int_{-\pi}^{\pi} \mathrm{tr}\, \Sigma_u^{PC}(\lambda) d\lambda = \frac{1}{2\pi} \sum_{k=r+1}^{n} \int_{-\pi}^{\pi} \rho_k(\lambda) d\lambda.$$

Therefore

$$0 \le \frac{\mathbf{E}(z_t^{PC})' z_t^{PC}}{\mathbf{E} x_t' x_t} = \frac{\sum_{k=1}^{r} \int_{-\pi}^{\pi} \rho_k(\lambda) d\lambda}{\sum_{k=1}^{n} \int_{-\pi}^{\pi} \rho_k(\lambda) d\lambda} \le 1 \tag{10.10}$$

is a measure of the fraction of the total variation of the process (x_t) explained by the principal components. Note that there are no restrictions on the number of factors $1 \le r < n$. So the user has to make a compromise between the fit of the model (see (10.10)) and the complexity of the model (measured by the number of factors).

The filter that describes the process of latent variables as a linear function of the observed process is either two-sided or static, whereby the static case is highly non-generic. In particular, this implies that the dynamic PCA model is not directly suited for prediction of the observed variables, since the latent variables also depend on future observations.

The dynamic PCA model, in general, gives non-rational spectral densities for the factors, latent variables and errors. This holds even in the case, when Σ_x is rational, since eigenvalues and eigenvectors are *not* rational functions of the respective matrix elements.

The estimation of DPCA from a given time series is quite demanding, which may be the reason that DPCA is not very popular, despite the many theoretical merits.

10.3   Generalized Dynamic Factor Models (GDFM)

As has been stated above, the noise is called strictly idiosyncratic if the spectral density Σ_u is diagonal. In the static case, where in addition (f_t) and (u_t) are assumed to be white noise, we have

$$\Gamma_x = \Gamma_z + \Gamma_u = L\Gamma_f L' + \Gamma_u. \tag{10.11}$$

Here Γ_x, Γ_z, Γ_f and Γ_u denote the covariance matrices of x_t, $z_t = Lf_t$, f_t and of u_t respectively. For given Γ_x, every decomposition as on the right-hand side of (10.11), where Γ_z is positive semidefinite and of rank $q < n$ and where Γ_u is positive semidefinite and diagonal, corresponds to a *static factor analysis model*.

For the dynamic case, an analogous decomposition is

$$\Sigma_x(\lambda) = \Sigma_z(\lambda) + \Sigma_u(\lambda) \tag{10.12}$$

where Σ_z is a singular and Σ_u is a diagonal spectral density. An analysis of the corresponding identifiability problem is given in Scherrer and Deistler (1998). Note that the assumption, that Σ_u is diagonal, is rather restrictive and the analysis of (10.12) is rather involved. This is the reason why, for high dimensional time series, generalized dynamic factor model are mostly used. In the following, the assumption of *strict idiosyncratic* noise is replaced by the assumption of *weak idiosyncratic* noise.

We start with a simple example. Let

$$x_t^n = \begin{pmatrix} x_{1t} \\ x_{2t} \\ \vdots \\ x_{nt} \end{pmatrix} = \begin{pmatrix} 1 \\ 1 \\ \vdots \\ 1 \end{pmatrix} v_t + \begin{pmatrix} u_{1t} \\ u_{2t} \\ \vdots \\ u_{nt} \end{pmatrix} \tag{10.13}$$

where $(v_t) \sim WN(\sigma^2)$ is (scalar) white noise and $(u_{it}) \sim WN(\sigma_i^2)$, $i = 1, \ldots, n$ are (scalar) white noise processes which are "weakly correlated" to each other in the sense that $|\mathbf{E}u_{it}u_{jt}| \le c < \infty$ for $|i - j| \le 1$ and $\mathbf{E}u_{it}u_{jt} = 0$ for $|i - j| > 1$). Then, according to a law of large numbers, for $n \to \infty$, the mean of the observed variables

$$\frac{1}{n} \sum_{i=1}^{n} x_{it} = v_t + \frac{1}{n} \sum_{i=1}^{n} u_{it} \xrightarrow[n \to \infty]{} v_t \tag{10.14}$$

converges (in mean square sense) to the factor v_t. This means that in a high dimensional scenario, it is possible to estimate the underlying factor by a suitable average of the observed variables. This is possible, since the noise components are only "weakly correlated" to each other and hence they may be eliminated by averaging. On the other hand, the latent variables are strongly related to each other.

Generalizing this idea leads to the following setting (see Chamberlain (1983), Chamberlain and Rothschild (1983) for the static case and Forni and Lippi (2001) for the dynamic case): We consider an underlying process with infinite cross-sectional dimension $(x_{it} \mid i \in \mathbb{N}, t \in \mathbb{Z})$ and we assume that the observed time series is generated by the finite-dimensional (sub-) process $(x_t^n = (x_{1t}, \ldots, x_{nt})' \mid t \in \mathbb{Z})$. In the following the limiting behavior for $n \to \infty$ is considered. This asymptotic analysis is justified, since we are mainly interested in high-dimensional time series. We assume that the observed variables may be decomposed as $x_{it} = z_{it} + u_{it}$, or equivalently, for each $n \in \mathbb{N}$,

$$x_t^n = z_t^n + u_t^n \tag{10.15}$$

where z_t^n and u_t^n are defined analogously to x_t^n. This decomposition, together with the assumptions below, gives a sequence of (nested) factor models and is called *generalized dynamic factor model* (GDFM). We in addition assume:

GFFM.0   For all $n \in \mathbb{N}$ the processes (z_t^n), (u_t^n) are stationary, orthogonal to each other ($\mathbf{E}z_{it}u_{js} = 0$ for all $i, j \in \mathbb{N}$ and $t, s \in \mathbb{Z}$) and they have spectral densities Σ_z^n and Σ_u^n respectively.

GDFM.1   For all $n \in \mathbb{N}$, Σ_z^n is a rational spectral density.

GDFM.2   There exist $q, n_0 \in \mathbb{N}$, such that the rank of Σ_z^n is equal to q for all $n \geq n_0$.

GDFM.3   There exist $m, n_1 \in \mathbb{N}$, such that the state dimension of a minimal state space system corresponding to Σ_z^n is equal to m for all $n \geq n_1$ onwards.

Clearly w.l.o.g. we may assume that $n_0 = n_1$. Next we define *weak* and *strong* *dependence*. We use $\rho_{z,s}^n(\lambda)$ and $\rho_{u,s}^n(\lambda)$ to denote the s-th largest eigenvalue of the spectral densities $\Sigma_z^n(\lambda)$ and $\Sigma_u^n(\lambda)$ respectively.

GDFM.4   (weak cross-sectional dependence of noise): The eigenvalues $\rho_{u,1}^n(\lambda)$ are uniformly bounded, i.e. there exists a $B > 0$ such that $\rho_{u,1}^n(\lambda) \leq B$ holds for all $\lambda \in [-\pi, \pi]$ and $n \in \mathbb{N}$.

GDFM.5   (strong cross-sectional dependence of the latent variables): The eigenvalues $\rho_{z,q}^n(\lambda)$ diverge as $n \to \infty$ for almost every $\lambda \in [-\pi, \pi]$.

These assumptions need some motivations and explanations.

First note that the vectors x_t^n, $n = 1, 2, \ldots$ are nested in the sense that the vector x_t^{n+1} is formed from x_t^n by adding an additional component, i.e. $x_t^{n+1} = ((x_t^n)', x_{n+1,t})'$. The same statement holds for the vectors z_t^n and u_t^n. This implies immediately that the rank of the covariance matrix $\Gamma_z^n = \mathbf{E}z_t^n(z_t^n)'$, the rank of the spectral density Σ_z^n and the dimension of a corresponding minimal state space representation are monotonically increasing with the dimension n. However, due to assumptions GDFM.0–GDFM.3 the rank of Σ_z and the minimal state dimension are constant from $n \geq n_0$ onwards. Clearly, the rank of the covariance matrix $\Gamma_z^n = \mathbf{E}z_t^n(z_t^n)'$ is bounded by the sum of the minimal state dimension and the rank of the spectral density, i.e. by $m + q$, see, e.g. (10.5) and (10.6). Thus we have shown that for $n \geq n_0$ the rank of the covariance matrix Γ_z^n is equal to some constant $q \leq r \leq m + q$. This implies, in particular, that all components $z_{it}, i \in \mathbb{N}$

are contained in the linear span $\text{sp}\{z_{it} \mid i = 1, \ldots, n_0\}$. Now, we determine a basis for this span. To this end, we factorize the covariance matrix $\Gamma_z^{n_0} = \mathbf{E} z_t^{n_0} (z_t^{n_0})'$ as $\Gamma_z^{n_0} = L^{n_0} (L^{n_0})'$, $L^{n_0} \in \mathbb{R}^{n_0 \times r}$ and define

$$f_t = ((L^{n_0})' L^{n_0})^{-1} (L^{n_0})' z_t^{n_0}. \tag{10.16}$$

It is clear that the components f_{it}, $i = 1, \ldots, r$ form the desired basis. Hence we have shown that

$$\begin{aligned} z_{it} &= (l_{i1}, \ldots, l_{ir}) f_t \\ z_t^n &= L^n f_t \end{aligned} \tag{10.17}$$

where the i-th row of the matrix $L^n \in \mathbb{R}^{n \times r}$ is given by $(l_{i1}, \ldots, l_{ir})$. Note that the (static) factors f_t (and the rows of the static factor loading matrix L^n) do *not* depend on n! The static factors f_t are unique up to a transformation with an orthogonal matrix and $\mathbf{E} f_t f_t' = I_r$ holds. Furthermore note that the latent variables and the factors are connected by a static, linear transformation, hence the (optimal) prediction of the latent variables may be based on the prediction of the (static) factors. We will come back to this issue at the end of this section.

Next we consider assumption GDFM.4. This assumption implies that the maximum eigenvalue of the covariance matrices $\Gamma_u^n = \mathbf{E} u_t^n (u_t^n)'$ is bounded. In other words, there exists a constant, $\bar{\gamma}$ say, such that

$$\Gamma_u^n \leq \bar{\gamma} I_n \quad \text{for all } n \in \mathbb{N}. \tag{10.18}$$

We call a sequence $(a^n \in \mathbb{R}^{1 \times n})_n$ such that $a^n (a^n)' \to 0$ for $n \to \infty$ an averaging sequence. From (10.18) we see that $a^n u_t^n$ converges to zero for $n \to \infty$ for all averaging sequences. Here, as always in this section, we consider convergence of random variables in the mean square sense. The condition (10.18), respectively the condition that the noise is annihilated by all averaging sequences, implies that the noise components only may by "weakly correlated". Consider a simple case where $\gamma_{ij}^u = \mathbf{E} u_{it} u_{jt} \geq 0$ holds for all i, j. A basic averaging sequence is $a^n = n^{-1} (1, \ldots, 1) \in \mathbb{R}^{1 \times n}$. Then $a^n u_t^n \to 0$ is equivalent to $\frac{1}{n^2} \sum_{ij=1}^n \gamma_{ij}^u \to 0$; hence, in particular, the case $\gamma_{ij}^u \geq c > 0$ for all i, j is not allowed.

If a^n is an averaging sequence and if $(a^n x_t^n)$ converges to a nonzero limit, then the limit has to be an element of the Hilbert space spanned by the components of the (static) factor f_t. Here the assumption GDFM.3 comes into play. However, in order to simplify the discussion, let us assume that

GDFM.3'   the r-th eigenvalue of the covariance matrix Γ_z^n converges to infinity.

Note that, by the above discussion, we know that $\Gamma_z^n = L^n (L^n)'$ and hence this condition is equivalent to the condition that the smallest eigenvalue of the matrix $(L^n)'(L^n) \in \mathbb{R}^{r \times r}$ diverges, i.e.

$$(L^n)'(L^n) \geq k(n) I_r \quad \text{and} \quad k(n) \to \infty.$$

The inequality above means that the factors have a "non-vanishing influence" on "infinitely many" latent variables. In this sense, the components of the latent variables are strongly correlated. If we know the factor loadings, then the factors may be estimated by a "denoising procedure" as follows. We define $A^n = ((L^n)'(L^n))^{-1}(L^n)'$ and note that $A^n(A^n)' = ((L^n)'(L^n))^{-1} \le (k(n))^{-1}I_q \to 0 \in \mathbb{R}^{r \times r}$. Thus each row of A^n is an averaging sequence and

$$A^n x_t^n = \underbrace{A^n L^n f_t}_{=f_t} + \underbrace{A^n u_t^n}_{\to 0} \xrightarrow[n \to \infty]{} f_t$$

as desired. The procedure described above is not a feasible estimation procedure for the factors f_t, because, e.g. the factor loading matrix L^n is not known. However, it can be shown that static PCA yields a feasible and consistent estimation procedure for the factors and the latent variables. It is shown in Chamberlain (1983), Chamberlain and Rothschild (1983), that under the above assumptions, the first r static principal components $(f_{1t}^{n,PC}, f_{2t}^{n,PC}, \ldots, f_{rt}^{n,PC})$, computed from the covariance matrix

$$\Gamma_x^n = \mathbf{E} x_t^n (x_t^n)' = L^n(L^n)' + \Gamma_u^n \tag{10.19}$$

are consistent "estimates" for the static factors f_t, in the sense, that there exists a sequence of matrices $O_n \in \mathbb{R}^{r \times r}$ such that

$$O_n(f_{1t}^{n,PC}, f_{2t}^{n,PC}, \ldots, f_{rt}^{n,PC})' \xrightarrow[n \to \infty]{} f_t.$$

A heuristic rational for this claim is, that the noise covariance Γ_u^n is bounded, whereas the covariance matrix $L^n(L^n)'$ of the latent variables is unbounded (in the sense GDFM.3'). Hence the "signal-to-noise ratio" in the decomposition $x_t = L^n f_t + u_t^n$ gets better and better with increasing n, and hence PCA yields better and better estimates for the factors f_t, since it is especially designed for such a situation. It can be shown, that this procedure works, even when the population covariance Γ_x^n is replaced by a consistent estimate, such as the sample covariance matrix. In this way, we obtain a proper estimation procedure.

This discussion has been substantially simplified using the assumptions GDFM.1–GDFM.3, since these assumptions lead to static factor loadings. These assumptions are also often used in practical applications. However, we should note that one may generalize the discussion to a truly dynamic version, as has been done in Forni and Lippi (2001). A dynamic averaging sequence is a sequence of (row functions) $a^n: [-\pi, \pi] \longrightarrow \mathbb{C}^{1 \times n}$ with $\int_{-\pi}^{\pi} a^n(\lambda)(a^n(\lambda))^* d\lambda \to 0$ for $n \to \infty$. If the noise (u_t^n) satisfies GDFM.4, then a^n defines a linear, dynamic transformation of the noise process $\Phi_{u^n}^{-1}(a^n)$ and the sequence of these transformations converges to zero. The condition GDFM.5 (together with the existence of a spectral density $\Sigma_z^n(\lambda)$ which has rank q μ-a.e.) implies that z_t^n has a Wold representation of the form $z_t^n = L^n(\mathbf{L})(\epsilon_t)$, where $(\epsilon_t) \sim \mathrm{WN}(I_q)$. Note that the white noise process (ϵ_t) and the rows of the filter $L^n(\mathbf{L})$ are independent of n. This yields a sequence of (nested) factor models $(x_t^n) = L^n(\mathbf{L})(\epsilon_t) + (u_t^n)$, with a q-dimensional dynamic factor process (ϵ_t). Furthermore, it can be shown that one may estimate the dynamic factors (and the latent variables) by suitable dynamic averaging sequences. In particular, it can be shown that dynamic principal component analysis may be used to this end.

As has been stated above, forecasting of (z_t) can be done by forecasting of (f_t). By GDFM.1–GDFM.3, (10.16) and (10.17) (f_t) has a rational spectral density of rank q. Thus (f_t) may be modeled by an ARMA system

$$\underline{a}(L)(f_t) = \underline{b}(L)(\epsilon_t) \tag{10.20}$$

satisfying the stability and the minimum phase assumptions and where $(\epsilon_t) \sim$ WN(I_q) is q-dimensional white noise. Equivalently, we also may use a state space system for (f_t)

$$s_{t+1} = As_t + B\epsilon_t$$
$$f_t = Cs_t + D\epsilon_t.$$

From (10.17), we then immediately obtain a state space representation for the latent variables

$$s_{t+1} = As_t + B\epsilon_t$$
$$z_t^n = L^n f_t = L^n Cs_t + L^n D\epsilon_t$$

Therefore $(D\epsilon_t)$ is the innovation process of (f_t) and $(L^n D\epsilon_t)$ is the innovation process of (z_t^n). Furthermore, ϵ_t are dynamic factors with minimal dimension q.

For many applications, the dimension r of minimal static factors f_t is larger than the dimension q of minimal dynamic factors ϵ_t. In this case, $\underline{b}(z)$ is a "tall" matrix. Now, "generically" in the parameter space for $q < r$, $\underline{b}(z)$ has no zeros, as "typically" the zeros of the determinant of a given $q \times q$ submatrix of $\underline{b}(z)$ are "compensated" by determinants of other $q \times q$ submatrices. If $\underline{b}(z)$ is a zeroless polynomial matrix, then its Smith(-McMillan) form (4.22) is of the form

$$\underline{b}(z) = u(z) \begin{pmatrix} I_q \\ 0 \end{pmatrix} v(z)$$

where $u(z)$ and $v(z)$ are unimodular polynomial matrices. We may extend $\underline{b}$ to an unimodular matrix

$$\tilde{\underline{b}}(z) = u(z) \begin{pmatrix} I_r & 0 \\ 0 & I_{r-q} \end{pmatrix} \begin{pmatrix} v(z) & 0 \\ 0 & I_{r-q} \end{pmatrix}$$

and write (10.20) as $\underline{a}(L)(f_t) = \tilde{\underline{b}}(L) \begin{pmatrix} \epsilon_t \\ 0 \end{pmatrix}$. As $\tilde{\underline{b}}(z)$ is unimodular, then we obtain

$$\tilde{\underline{b}}^{-1}(L)\underline{a}(L)(f_t) = \begin{pmatrix} \epsilon_t \\ 0 \end{pmatrix},$$

i.e. (f_t) is generically an AR process; for more details see Anderson and Deistler (2008). As has been mentioned already, AR systems are easier to estimate. The AR system here, i.e. for $q < r$, has singular innovation variance. Such systems have been discussed in Chen et al. (2011).

Based on the discussion above, in many cases, forecasting models for high dimensional times series are obtained by the following procedure (for more details see, e.g. Forni et al. (2009) and Stock and Watson (2016)):

(1) Estimate the static factors f_t and the noise u_t via static PCA for the observed time series.
(2) The static factors are modeled by an AR model. Estimate q from the innovation variance, see, e.g. Bai and Ng (2002).
(3) Estimate univariate AR models for the components u_{it} of the noise.
(4) Finally the forecast for x_{t+h} is obtained by adding the forecast of the latent variables (obtained by the forecast of the static factors and the loading matrix) and the forecasts for the univariate noise components.

Of course this is only one of the many possibilities for modeling and forecasting with (generalized) dynamic factor models.

References

B.D.O. Anderson, M. Deistler, Properties of Zero-free transfer function matrices. SICE J. Control Meas. Syst. Integr. **1**(4), 284–292 (2008). (July)

J. Bai, Inferential theory for factor models of large dimension. Econometrica **71**(1), 135–171 (2003). ISSN 1468-0262. https://doi.org/10.1111/1468-0262.00392

J. Bai, S. Ng, Determining the number of factors in approximate factor models. Econometrica **70**(1), 191–221 (2002). ISSN 0012-9682. https://doi.org/10.1111/1468-0262.00273

D.R. Brillinger, *Time Series: Data Analysis and Theory*. Classics in Applied Mathematics. Society for Industrial and Applied Mathematics, 2001 (Originally Published, Holden-Day, 1981). https://doi.org/10.1137/1.9780898719246

C. Burt, Experimental tests of general intelligence. British J. Psychol. 1904–1920, **3**(1–2), 94–177 (1909). https://doi.org/10.1111/j.2044-8295.1909.tb00197.x. https://bpspsychub.onlinelibrary.wiley.com/doi/abs/10.1111/j.2044-8295.1909.tb00197.x

G. Chamberlain, Funds, factors, and diversification in arbitrage pricing models. Econometrica **51**(5), 1305–1323 (1983). (Sept.)

G. Chamberlain, M. Rothschild, Arbitrage, factor structure, and mean-variance analysis on large asset markets. Econometrica **51**(5), 1281–1304 (1983). (Sept.)

W. Chen, B.D. Anderson, M. Deistler, A. Filler, Solutions of Yule-Walker equations for singular AR processes. J. Time Ser. Anal. **32**(5), 531–538 (2011). ISSN 1467-9892. https://doi.org/10.1111/j.1467-9892.2010.00711.x

R. Diversi, R. Guidorzi, U. Soverini, Maximum likelihood identification of noisy input-output models. Automatica **43**(3), 464–472 (2007). ISSN 0005-1098. https://doi.org/10.1016/j.automatica.2006.09.009

C. Doz, D. Giannone, L. Reichlin, A two-step estimator for large approximate dynamic factor models based on Kalman filtering. J. Economet. **164**(1), 188–205 (2011). ISSN 0304-4076.

https://doi.org/10.1016/j.jeconom.2011.02.012. https://www.sciencedirect.com/science/article/pii/S030440761100039X. Annals Issue on Forecasting

M. Forni, M. Lippi, The generalized dynamic factor model: representation theory. Economet. Theory **17**, 1113–1141, JEL Classif. C13, C **33**, C43 (2001)

M. Forni, M. Hallin, M. Lippi, L. Reichlin, The generalized dynamic-factor model: identification and estimation. Rev. Econ. Stat. **82**(4), 540–554 (2000). (November)

M. Forni, D. Giannone, M. Lippi, L. Reichlin, Opening the black box: structural factor models versus structural VARs. Economet. Theory **25**, 1319–1347 (2009)

J.F. Geweke, The dynamic factor analysis of economic time series, in *Latent Variables in Socioeconomic Models*. ed. by D. Aigner, A. Goldberger (North Holland, Amsterdam, 1977)

M. Hallin, M. Lippi, M. Barigozzi, M. Forni, P. Zaffaroni, *Time Series in High Dimensions: the General Dynamic Factor Model* (World Scientific, NJ, 2020). 9813278005

D.N. Lawley, A.E. Maxwell, *Factor Analysis as a Statistical Method*, 2nd edn. (Butterworth & Co., 1971)

M. Lippi, M. Deistler, B. Anderson, High-Dimensional dynamic factor models: a selective survey and lines of future research. To appear in: *Econometrics and Statistics* (2022)

P. Poncela, E. Ruiz, K. Miranda, Factor extraction using Kalman filter and smoothing: this is not just another survey. Int. J. Forecast. **37**(4), 1399–1425 (2021). ISSN 0169-2070. https://doi.org/10.1016/j.ijforecast.2021.01.027. https://www.sciencedirect.com/science/article/pii/S0169207021000273

T.J. Sargent, C.A. Sims, Business cycle modeling without pretending to have too much a priori economic theory, in *New Methods in Business Cycle Research: Proceedings from a Conference*. ed. by C.A. Sims (Federal Reserve Bank of Minneapolis, Minneapolis, 1977), pp.45–109. (Jan.)

W. Scherrer, M. Deistler, A structure theory for linear dynamic errors-in-variables models. SIAM J. Control Optim. **36**(6), 2148–2175 (1998). (Nov.)

C. Spearman, General intelligence, objectively determined and measured. Am. J. Psych. **15**, 201–293 (1904)

J.H. Stock, M.W. Watson, Forecasting using principal components from a large number of predictors. J. Am. Stat. Assoc. **97**(460), 1167–1179 (2002)

J.H. Stock, M.W. Watson, Dynamic factor models, factor-augmented vector autoregressions, and structural vector autoregressions in macroeconomics, in *Handbook of Macroeconomics*, vol. 2, ed. by J.B. Taylor, H. Uhlig (Elsevier, Amsterdam, 2016), pp. 415–525

ARCH and GARCH Models

11

In finance data one often observes so-called volatility clustering, i.e. periods with relatively high volatility and periods with low volatility occur. This is an indication that the (conditional) variance is dependent on past observations. The most common models for the conditional variance are (G)ARCH models and stochastic volatility (SV) models. Both are often used to model financial data, e.g. asset returns. Here we only discuss the GARCH case, since combining AR/ARMA Models with GARCH innovations provides an easy way to model jointly the conditional mean and the conditional variance. A central result in this section is a necessary and sufficient condition for stationary solutions of (G)ARCH systems.

Consider a regular, stationary process (x_t) with a rational spectral density (of full rank). As has been shown above, such a process may be represented by a state space model (in innovation form) or equivalently by a stable and minimum phase ARMA model

$$x_t = a_1 x_{t-1} + \cdots + a_p x_{t-p} + \epsilon_t + b_1 \epsilon_{t-1} + \cdots + b_q \epsilon_{t-q}.$$

Here in addition, we assume that the innovation process (ϵ_t) is a *martingale difference sequence (MDS)*, i.e. we require that the conditional expectation of ϵ_{t+1} given the past is zero: $\mathbf{E}(\epsilon_{t+1} \mid \epsilon_s, \ s \leq t) = 0$. Denote the sigma algebra generated by the past of a process (u_t) by $\mathcal{F}_t(u) = \sigma\{u_s \mid s \leq t\}$. Then we have

$$\mathbf{E}(\epsilon_{t+1} \mid \mathcal{F}_t(\epsilon)) = 0.$$

Note also that $\mathcal{F}_t(\epsilon) = \sigma\{\epsilon_s \mid s \leq t\} = \sigma\{x_s \mid s \leq t\} = \mathcal{F}_t(x)$, since (ϵ_t) is the innovation process and hence $\mathbb{H}_t(x) = \mathbb{H}_t(\epsilon)$. Now, it is immediate that

$$\mathbf{E}(x_{t+1} \mid \mathcal{F}_t(x)) = a_1 x_t + \cdots + a_p x_{t+1-p} + b_1 \epsilon_t + \cdots + b_q \epsilon_{t+1-q} = \mathrm{P}_{\mathbb{H}_t(x)} x_{t+1} = \hat{x}_{t,1}$$

As has been noted above, we see that (given this MDS condition) the conditional mean of the observed variables given the past is determined by the ARMA model.

© Springer Nature Switzerland AG 2022
M. Deistler and W. Scherrer, *Time Series Models*, Lecture Notes in Statistics 224,
https://doi.org/10.1007/978-3-031-13213-1_11

Furthermore, it follows that the *linear, least squares* forecast is the overall best forecast in terms of the mean squared error.

In many applications, not only the forecast (conditional mean) is of interest but it is also important to have a measure for the quality of the forecast, in particular, the variance of the prediction errors is of interest. Of course, due to stationarity, the variance of the observed variables, as well as the variance of the prediction errors (innovations) $\epsilon_{t+1} = x_{t+1} - \hat{x}_{t,1}$, is constant (independent of time t). However, the conditional variance

$$\Sigma_{t+1} := \mathbf{Var}(x_{t+1} \,|\, \mathcal{F}_t(x)) = \mathbf{Var}(\epsilon_{t+1} \,|\, \mathcal{F}_t(\epsilon))$$

may be *non constant*. This conditional variance, e.g. may be used for risk analysis in finance.

In the following, we will discuss the most common models for conditional variances: The ARCH (AutoRegressive Conditional Heteroscedasticity) model was introduced in the scalar case by Engle (1982) and then generalized to the GARCH (Generalized ARCH) model by Bollerslev (1986). We will also present some of the many multivariate GARCH (MGARCH) models, in particular the VECH model Bollerslev et al. (1988) and the BEKK model Baba et al. (1991). These *non-linear* models, in a certain sense, are outside the main scope of the book, which mostly deals with *linear* time series models. Therefore, this section only provides some of the basic ideas related to (M)GARCH models.

11.1   Models for the Conditional Variance

We start with a scalar (not necessarily stationary) process (ϵ_t).

Definition 11.1 A scalar process (ϵ_t) is called *GARCH(p, q) process (generalized autoregressive conditionally heteroskedasticity process of order (p, q))*, if the conditional mean is zero ($\mathbf{E}(\epsilon_t \,|\, \mathcal{F}_{t-1}(\epsilon)) = 0$) and the conditional variance

$$s_t^2 = \mathbf{Var}(\epsilon_t \,|\, \mathcal{F}_{t-1}(\epsilon)) = \mathbf{E}(\epsilon_t^2 \,|\, \mathcal{F}_{t-1}(\epsilon))$$

satisfies the following equation

$$s_t^2 = \gamma + \alpha_1 \epsilon_{t-1}^2 + \cdots + \alpha_q \epsilon_{t-q}^2 + \alpha_{1+q} s_{t-1}^2 + \cdots + \alpha_{q+p} s_{t-p}^2 \qquad (11.1)$$

with $\gamma, \alpha_i \in \mathbb{R}, q > 0, \alpha_q \neq 0$ and $\alpha_{q+p} \neq 0$ for $p > 0$. If $p = 0$ then the process is called ARCH(q) *process*. (Note that the case $q = 0$ does not make sense.)

In the ARCH model the conditional variance is a function of the past q squared observations. Since, as we will discuss below, the coefficients $\alpha_i \geq 0$ are non-negative, large (squared) observations will lead to a large conditional variance and hence there is a tendency, that also the next observation is large (in absolute value). This effect, called *volatility clustering*, is often observed in financial data. Adding the "GARCH terms" $\alpha_{i+q} s_{t-i}^2$ leads to an "infinite memory" in the sense that the conditional variance now depends on observations in the remote past of the process. (However, a stationary solution only exists, if this dependency fades away sufficiently fast.)

In the non-stationary case, one often considers the process only for ($t \in \mathbb{N}$). In this case, the above GARCH Eq. (11.1) has to hold for all $t \in \mathbb{N}$ with suitable initial conditions $\epsilon_{1-q}, \ldots, \epsilon_0$ and $s_{1-p}^2, \ldots, s_0^2$.

The GARCH model describes the conditional mean and variance of the observed process. If, in addition, the conditional distribution is prescribed, e.g.

$$(\epsilon_t \,|\, \mathcal{F}_{t-1}(\epsilon)) \sim N(0, s_t^2),$$

then maximum likelihood methods may be used for estimation of the parameters.

Up to now, no restrictions on the parameters have been imposed. However, since s_t^2 is a conditional variance, we have to assure that the right-hand side of Eq. (11.1) is always positive (or at least non-negative). Therefore one assumes[1]

$$\gamma > 0, \; \alpha_i \geq 0.$$

The conditions for the existence (and uniqueness) of a (weakly) stationary solution of a GARCH model will be discussed at the end of this section.

Now we want to carry over the construction of scalar GARCH process to the multivariate case (MGARCH processes/models). To this end, we need some notation. The vec operator stacks the columns of a matrix into a (large) column vector. Symmetric matrices are determined by the entries on and below the main diagonal, hence the "half-vectorisation" operator vech is introduced, which stacks the elements on and below the main diagonal into a vector. The reverse operators (which reconstruct the matrix from the vectorized version) are denoted with mat and math, respectively. If $u \in \mathbb{R}^n$ is a (column) vector, then $\text{diag}(u) \in \mathbb{R}^{n \times n}$ is the corresponding diagonal matrix. Furthermore $A \otimes B$ denotes the Kronecker product of the matrices A and B and if A, B have the same dimension, then we may compute the Hadamard product $A \odot B$ of A and B. (The Hadamard product is the elementwise product of the two matrices.).

The most general specification of multivariate GARCH models, which will be discussed here, is the so-called VEC model, introduced by Bollerslev et al. (1988)

$$\text{vec}(S_t) = V_0 + \sum_{i=1}^{q} V_i \text{vec}(\epsilon_{t-i} \epsilon_{t-i}') + \sum_{i=1}^{p} V_{i+q} \text{vec}(S_{t-i}) \qquad (11.2)$$

where $S_t = \text{Var}(\epsilon_t \,|\, \mathcal{F}_{t-1}(\epsilon))$ is the conditional variance (matrix). Note that the matrices $V_i, i = 1, \ldots, p + q$ are of dimension $n^2 \times n^2$ and therefore the number

---

[1] This assumption is sufficient but not necessary to guarantee that the recursions (11.1) always generate non-negative outputs (s_t^2) for arbitrary (non-negative) inputs (ϵ_t^2). However, the sufficient and necessary conditions are much harder to handle.

of parameters of this general specification quickly explodes with the dimension n of the observed process (ϵ_t). Furthermore, it is hard to come up with suitable (simple to check and not too restrictive) conditions on the parameters which ensure that the right-hand side of the above Eq. (11.2) is positive (semi-) definite for arbitrary ϵ_{t-i} and $S_{t-i} \geq 0$. A VEC parameter matrix V is called *admissible* if

$$\text{mat}(V \text{vec}(ee')) \geq 0 \text{ for all } e \in \mathbb{R}^n$$

holds. Clearly, the above VEC model yields a positive definite conditional variance S_t for arbitrary ϵ_{t-i} and $S_{t-i} \geq 0$, if $\text{mat}(V_0)$ is positive definite and if all V_i's are admissible.

Exercise 11.2 Prove that

$$\text{mat}(V \text{vec}(ee')) \geq 0 \,\forall e \in \mathbb{R}^n \iff \text{mat}(V \text{vec}(S)) \geq 0 \,\forall S \in \mathbb{R}^{n \times n}, \; S \geq 0.$$

The conditional variances S_{t-i} and $(\epsilon_{t-i} \epsilon_{t-i}')$ are (by definition) symmetric matrices and therefore it suffices to consider the entries on and below the main diagonal. If we explicitly consider this constraint and neglect redundant terms, we obtain the so-called VECH model:

$$\text{vech}(S_t) = \tilde{V}_0 + \sum_{i=1}^q \tilde{V}_i \text{vech}(\epsilon_{t-i} \epsilon_{t-i}') + \sum_{i=1}^p \tilde{V}_{i+q} \text{vech}(S_{t-i}). \tag{11.3}$$

The parameter matrices $\tilde{V}_i$ of the VECH model are of dimension $(n(n+1)/2) \times (n(n+1)/2)$. The number of parameters is still of order $O(n^4)$, which means that, except for the bivariate case $n = 2$, such a general VECH specification is rarely used in practice. Of course this re-parametrization also does not help in finding simple constraints which guarantee that the model always yields positive definite matrices S_t.

Because of these difficulties a number of alternative (restricted) models have been proposed in the literature:

The diagonal VEC model (DVEC) assumes that the parameter matrices V_i, $i = 1, \ldots, p+q$ are diagonal matrices. This means that the conditional covariance, $S_{ij,t}$ say, only depends on "its own past" $S_{ij,s}$, $s < t$ and the respective products $\epsilon_{is} \epsilon_{js}$, $s < t$. This seems a reasonable assumption for many applications and it substantially reduces the number of parameters (to $O(n^2)$) and the admissibility of the parameter matrices is rather simple to check. We leave this as an exercise to the reader:

Exercise 11.3 Show that a diagonal VEC model (with $V_i = \text{diag}(v_i), i = 1, \ldots, q$) may be written as

$$S_t = D_0 + \sum_{i=1}^q D_i \odot (\epsilon_{t-i} \epsilon_{t-i}') + \sum_{i=1}^p D_{i+q} \odot S_{t-i}$$

where $D_0 = \text{mat}(V_0)$ and $D_i = \text{mat}(v_i)$. Furthermore, prove that a diagonal VEC parameter matrix $V = \text{diag}(v)$ is admissible if and only if the matrix $\text{mat}(v)$ is positive semidefinite. Hint: For the last claim, one has to prove that for a square (symmetric) matrix $D \in \mathbb{R}^{n \times n}$, the matrix $(D \odot (ee')) \geq 0$ is positive semidefinite for all vectors $e \in \mathbb{R}^n$, if and only if D is positive semidefinite.

The BEKK model, named after the authors Baba et al. (1991), is a model of the form

$$S_t = B_0 B_0' + \sum_{i=1}^{q} \sum_{j=1}^{r} B_{i,j} \epsilon_{t-i} \epsilon_{t-i}' B_{i,j}' + \sum_{i=1}^{p} \sum_{j=1}^{r} B_{i+q,j} S_{t-i} B_{i+q,j}' \qquad (11.4)$$

with a lower diagonal matrix $B_0 \in \mathbb{R}^{n \times n}$ and matrices $B_{i,j} \in \mathbb{R}^{n \times n}$. Note that (for arbitrary square matrices S, $B_1, \ldots, B_r \in \mathbb{R}^{n \times n}$)

$$\text{vec}\left(\sum_{j=1}^{r} B_j S B_j' \right) = \left(\sum_{j=1}^{r} (B_j \otimes B_j) \right) \text{vec}(S) \qquad (11.5)$$

and therefore any BEKK model may be easily represented as VEC(H) model. It is also clear that admissibility is not an issue for BEKK models, since $\sum_{j=1}^{r} B_j S B_j'$ is positive semidefinite for arbitrary $S \geq 0$. In other words, the VEC parameter $V = \sum_{j=1}^{r} (B_j \otimes B_j)$ corresponding to a BEKK term is always admissible. The reverse question, whether an admissible VEC matrix may be represented by a BEKK is discussed in Scherrer and Ribarits (2007). In general, $r = n^2$ terms are needed to cover as many admissible VEC matrices as possible. For $n = 2$ every admissible VEC matrix may be represented by a BEKK term (with $r = n^2 = 4$). However for $n > 2$ there is a "thick" set of admissible VEC matrices which may not be cast as BEKK terms.

It should also be noted that the BEKK parameters are not identified without further restrictions. (E.g. note that any permutation of the matrices $B_1, \ldots, B_r$ of course gives the same result.) As has been stated above the most general model(s) need $r = n^2$ and hence the number of free parameters is $O(n^4)$ like for the VEC(H) model. In the following, we consider two important restricted classes of BEKK models, which are more parsimonious ($r < n^2$).

In the diagonal BEKK model the matrices $B_{i,j} = \text{diag}(b_{i,j}) \in \mathbb{R}^{n \times n}$, $i = 1, \ldots, q + p$, $j = 1, \ldots, r$ are diagonal matrices and (hence) $r \leq n$. This leads to a diagonal VEC model (see (11.5)). The following exercise shows that also the reverse statement holds, and thus diagonal BEKK models and (admissible) diagonal VEC models are equivalent.

Exercise 11.4 Show that an *admissible* DVEC parameter matrix $D \geq 0$ admits a diagonal BEKK representation, i.e. there exist diagonal matrices $B_j = \text{diag}(b_j)$, $j = 1, \ldots, r$ with $r < n$ such that

$$D \odot S = \sum_{j=1}^{r} B_j S B'_j \text{ for all } S \geq 0.$$

Note that a DVEC parameter matrix $D \in \mathbb{R}^{n \times n}$ is admissible if $(D \odot S) \geq 0$ holds for all $S \geq 0$. As has been shown in Exercise (11.3) this is equivalent to $D \geq 0$ and hence one may consider the eigenvalue decomposition of $D = \sum_{j=1}^{r} u_j \lambda_j u'_j = \sum_{j=1}^{r} b_j b'_j$.

A further reduction of parameters is obtained by the so-called factor GARCH model proposed in Engle et al. (1990). Consider a non-singular matrix $\Lambda = (\lambda_1, \dots, \lambda_n) \in \mathbb{R}^{n \times n}$ with columns λ_i and its inverse $\Lambda^{-1} = \Gamma' = (\gamma_1, \dots, \gamma_n)'$ with rows γ'_i. Furthermore the columns of Γ are normalized by $\|\gamma_i\| = 1$. The BEKK parameter matrices are rank one matrices of the form

$$B_{i,j} = \alpha_{i,j} \lambda_j \gamma'_j.$$

By construction it holds that

$$I_n = \Lambda \Gamma' = \sum_{i=1}^{n} \lambda_i \gamma'_i$$
$$I_n = \Gamma' \Lambda = (\gamma'_i \lambda_j)_{i,j=1,\dots,n} \qquad \text{and thus } \gamma'_i \lambda_i = 1 \text{ and } \gamma'_i \lambda_j = 0 \, \forall i \neq j.$$

This provides a factor decomposition of the observed process (ϵ_t) as

$$\epsilon_t = \sum_{k=1}^{r} \lambda_k \underbrace{\gamma'_k \epsilon_t}_{f_{kt}} + \underbrace{\sum_{k=r+1}^{n} \lambda_k \gamma'_k \epsilon_t}_{v_t} = \sum_{i=1}^{r} \lambda_k f_{kt} + v_t \qquad (11.6)$$

with $1 \leq r \leq n$ factors $f_{kt} = \gamma'_k \epsilon_t$, $k = 1, \dots, r$, factor loadings λ_k and a "noise" term $v_t = M \epsilon_t$ where $M = \sum_{k=r+1}^{n} \lambda_k \gamma'_k$. We now compute the respective conditional variances and covariances of these variables:

$$\mathbf{Var}(f_{kt} \mid \mathcal{F}_t(\epsilon)) = \gamma'_k S_t \gamma_k$$
$$= \gamma'_k B_0 B'_0 \gamma_k + \sum_{i=1}^{q} \alpha^2_{i,k} f^2_{k,t-i} + \sum_{i=1}^{p} \alpha^2_{i+q,k} \mathbf{Var}(f_{k,t-i} \mid \mathcal{F}_t(\epsilon))$$
$$\mathbf{Cov}(f_{kt}, f_{lt} \mid \mathcal{F}_t(\epsilon)) = \gamma'_k S_t \gamma_l$$
$$= \gamma'_k B_0 B'_0 \gamma_l = \mathbf{Cov}(f_{kt}, f_{lt}) \text{ for } k \neq l$$
$$\mathbf{Var}(v_t \mid \mathcal{F}_t(\epsilon)) = M S_t M'$$
$$= M B_0 B'_0 M' = \mathbf{Var}(v_t)$$
$$\mathbf{Cov}(v_t, f_{kt} \mid \mathcal{F}_t(\epsilon)) = M S_t \gamma_k$$
$$= M B_0 \gamma_k = \mathbf{Cov}(v_t, f_{kt}).$$

Thus, the factors (f_{kt}) are univariate GARCH processes. The conditional variance matrix of the noise (v_t) and the conditional covariance matrices are constant. For $r = n$ there is no noise term, and hence ϵ_t is a linear combination of scalar GARCH processes only.

It is an important question whether the VEC model (11.2) (and its variants) have a (strictly or weakly) stationary solution. If the VEC process (ϵ_t) is (weakly) stationary then (ϵ_t) is a white noise process. By elementary properties of the conditional mean, we get

$$\mathbf{E}\epsilon_t = \mathbf{E}(\mathbf{E}(\epsilon_t \mid \mathcal{F}_{t-1}(\epsilon))) = \mathbf{E}0 = 0$$
$$\mathbf{E}\epsilon_{t+k}\epsilon_t' = \mathbf{E}(\mathbf{E}\epsilon_{t+k}\epsilon_t' \mid \mathcal{F}_t(\epsilon))) = \mathbf{E}(\mathbf{E}\epsilon_{t+k} \mid \mathcal{F}_t(\epsilon))\epsilon_t') = \mathbf{E}0 = 0 \ \forall k > 0.$$

Furthermore, we see that

$$\Sigma := \mathbf{E}\epsilon_t\epsilon_t' = \mathbf{E}(\mathbf{E}(\epsilon_t\epsilon_t' \mid \mathcal{F}_{t-1}(\epsilon))) = \mathbf{E}S_t$$

and thus from the basic VEC Eq. (11.2) it follows that

$$\mathrm{vec}(\Sigma) = V_0 + \sum_{i=1}^{q} V_i \mathrm{vec}(\Sigma) + \sum_{i*1}^{p} V_{i+p} \mathrm{vec}(\Sigma)$$
$$= V_0 + \left(\sum_{i=1}^{p+q} V_i\right) \mathrm{vec}(\Sigma).$$

This means that (for all $k \geq 0$)

$$\Sigma = \mathrm{mat}(V_0) + \mathrm{mat}(V V_0) + \cdots + (\mathrm{mat}(V^k V_0) + \mathrm{mat}(V^{k+1}\mathrm{vec}(\Sigma)))$$

where $V = \left(\sum_{i=1}^{p+q} V_i\right)$. Note that V is admissible and hence all terms $\mathrm{mat}(V^s V_0)$ are positive semidefinite matrices. Thus the above equation for Σ has a positive definite (and finite) solution for all V_0, $\mathrm{mat}(V_0) > 0$ if and only if all the eigenvalues of V are less than one in modulus. It is shown in Engle and Kroner (1995) that this condition is also sufficient for the existence of a (weakly) stationary solution. In this case, the covariance matrix Σ of ϵ_t is given by

$$\Sigma = \mathrm{mat}\left((I - V)^{-1}V_0\right).$$

Clearly this discussion also holds for all models, which may be cast as VEC model, e.g. the corresponding condition for the BEKK model is that the eigenvalues of the matrix

$$\sum_{i=1}^{p+q}\sum_{j=1}^{r}(B_j \otimes B_j)$$

are less than one in modulus.

References

Y. Baba, R. F. Engle, D. F. Kraft, K.F. Kroner, Multivariate Simultaneous Generalised ARCH. Technical report, University of California, San Diego: Department of Economics, Discussion Paper No. 89–57 (1991)

T. Bollerslev, Generalised autoregressive heteroscedasticity. J. Economet. **31**, 307–27 (1986)

T. Bollerslev, R. Engle, J. Wooldridge, A capital asset pricing model with time varying covariance. J. Political Econ. **96**, 116–131 (1988)

R. Engle, Autoregressive conditional heteroscedasticity with estimates of the variance of the U.K. Inflation. Econometrica **50**, 987–1008 (1982)

R.F. Engle, K.F. Kroner, Mutlivariate simultaneous generalized ARCH. Economet. Theory **11**, 122–150 (1995)

R.F. Engle, V.K. Ng, M. Rothschild, Asset pricing with a factor-ARCH Covariance structure, empirical estimates for treasury bills. J. Economet. **45**(1–2), 213–237 (1990). ISSN 0304-4076. https://doi.org/10.1016/0304-4076(90)90099-F. http://www.sciencedirect.com/science/article/pii/030440769090099F

W. Scherrer, E. Ribarits, On the parametrization of multivariate GARCH models. Economet. Theory **23**, 464–484, 5 (2007). ISSN 1469-4360. https://doi.org/10.1017/S026646660707020X. http://journals.cambridge.org/article_S026646660707020X

Index

© Springer Nature Switzerland AG 2022
M. Deistler and W. Scherrer, *Time Series Models*, Lecture Notes in Statistics 224,
https://doi.org/10.1007/978-3-031-13213-1

Printed in the United States
by Baker & Taylor Publisher Services